THE PHYSICS BOOK

WORKBOOK

UNITS 3 4

Scott Adamson
Oliver Alini
Neil Champion
Tara Kuhn

The Physics Book Units 3 & 4 Workbook
1st Edition
Scott Adamson
Oliver Alini
Neil Champion
Tara Kuhn

Contributing authors: Gary Bass, Geoff Cody, Robert Farr, Suzanne Garr, Roger Walter and Kate Wilson
Publisher: Siobhan Moran and Rachel Ford
Project editor: Simon Tomlin
Editor: Gene Anderson-Conklin
Proofreader: Carly Slater
Permissions researcher: Corrina Gilbert
Cover image: istock/sakkmesterke
Cover design: Chris Starr (MakeWork)
Text design: Watershed Design
Production controller: Karen Young
Typeset by: MPS Limited

Acknowledgements
Physics 2019 v1.2 General Senior Syllabus © Queensland Curriculum & Assessment Authority.This syllabus forms part of a new senior assessment and tertiary entrance system in Queensland. Along with other senior syllabuses, it is still being refined in preparation for implementation in schools from 2019.For the most current syllabus versions and curriculum information please refer to the QCAA website https://www.qcaa.qld.edu.au/. The QCAA's permission does not imply permission to reproduce non-QCAA material. For permission to reproduce the material for which the QCAA owns copyright, please contact the QCAA, PO Box 307, Spring Hill Qld 4004; publishing@qcaa.qld.edu.au.

For product information and technology assistance,
in Australia call **1300 790 853**;
in New Zealand call **0800 449 725**

For permission to use material from this text or product, please email **aust.permissions@cengage.com**

ISBN 978 0 17 041264 3

Cengage Learning Australia
Level 7, 80 Dorcas Street
South Melbourne, Victoria Australia 3205

Cengage Learning New Zealand
Unit 4B Rosedale Office Park
331 Rosedale Road, Albany, North Shore 0632, NZ

For learning solutions, visit **cengage.com.au**

Printed in Singapore by 1010 Printing Group Limited.
1 2 3 4 5 6 7 22 21 20 19 18

CONTENTS

4 Circular motion 72

5 Gravitational force and field 91

6 Orbital motion 104

TOPIC 2: ELECTROMAGNETISM

UNIT 4 » REVOLUTIONS IN MODERN PHYSICS 151

TOPIC 1: SPECIAL RELATIVITY

TOPIC 2: QUANTUM THEORY

9780170412643

TOPIC 3: THE STANDARD MODEL

HOW TO USE THIS BOOK

Learning

The learning section is a summary of the key knowledge and skills. This summary can be used to create mind maps, to write short summaries and as a check list.

Revision

This section is a series of structured activities to help consolidate the knowledge and skills acquired in class.

Evaluation

The evaluation section is in the style of a practice exam to test and evaluate the acquisition of knowledge and skills.

Practice exam

A tear-out exam helps to facilitate preparing and practicing for external exams.

ABOUT THE AUTHORS

Scott Adamson

An experienced maths author, science reviewer, HOD Science and member of the QCAA Physics State Panel and QCAA Science LARG, Scott brings a wealth of knowledge in teaching physics, as well as teaching senior science, to the author team. Scott's dedication to teaching physics has enabled him to lead the *QScience* team in the development of this text.

Oliver Alini

Oliver is a dedicated senior physics teacher and assistant boarding house master. In addition to his experience in teaching physics, Oliver has also been a maths teacher and tutor.

Neil Champion

Neil was directly involved in writing the Australian Senior Physics Curriculum. Neil is an experienced physics author, leading the team on *Nelson Physics for the Australian Curriculum*. Neil has taught physics at both a high-school and a university level, including the physics teaching methodology.

Tara Kuhn

An experienced science teacher, Tara brings her enthusiasm for physics and maths to the writing team.

REVIEW TEAM

The following people have contributed to the review of the *QScience Physics* series: David Austin, Neil Gordon, Rebecca Delaney and Catherine Munro.

Some of the material in *The Physics Book Units 3 & 4* has been taken from or adapted from the following: *Nelson Physics for the Australian Curriculum Units 1 & 2* NelsonNet, written by Neil Champion, Geoff Cody, Robert Farr, Suzanne Garr and Roger Walter.

Nelson Physics for the Australian Curriculum Units 3 & 4 NelsonNet, written by Neil Champion, Kate Wilson, Rob Farr, Roger Walter and Gary Bass.

SYLLABUS REFERENCE GRID

UNITS AND TOPICS	*NELSON QSCIENCE PHYSICS UNITS 3 & 4*
UNIT THREE: GRAVITY AND ELECTROMAGNETISM	
Topic 1: Gravity and motion	
Vector analysis	Chapter 1
Projectile motion	Chapter 2
Inclined plane	Chapter 3
Circular motion	Chapter 4
Gravitational force and field	Chapter 5
Orbital motion	Chapter 6
Topic 2: Electromagnetism	
Electrostatics	Chapter 7
Magnetism	Chapter 8
Electromagnetic induction	Chapter 9
Electromagnetic radiation	Chapter 10
UNIT FOUR: REVOLUTIONS IN MODERN PHYSICS	
Topic 1: Special relativity	
Special relativity	Chapter 11
Topic 2: Quantum theory	
Quantum theory	Chapter 12
Topic 3: The Standard Model	
The Standard Model	Chapter 13
Particle interactions	Chapter 14

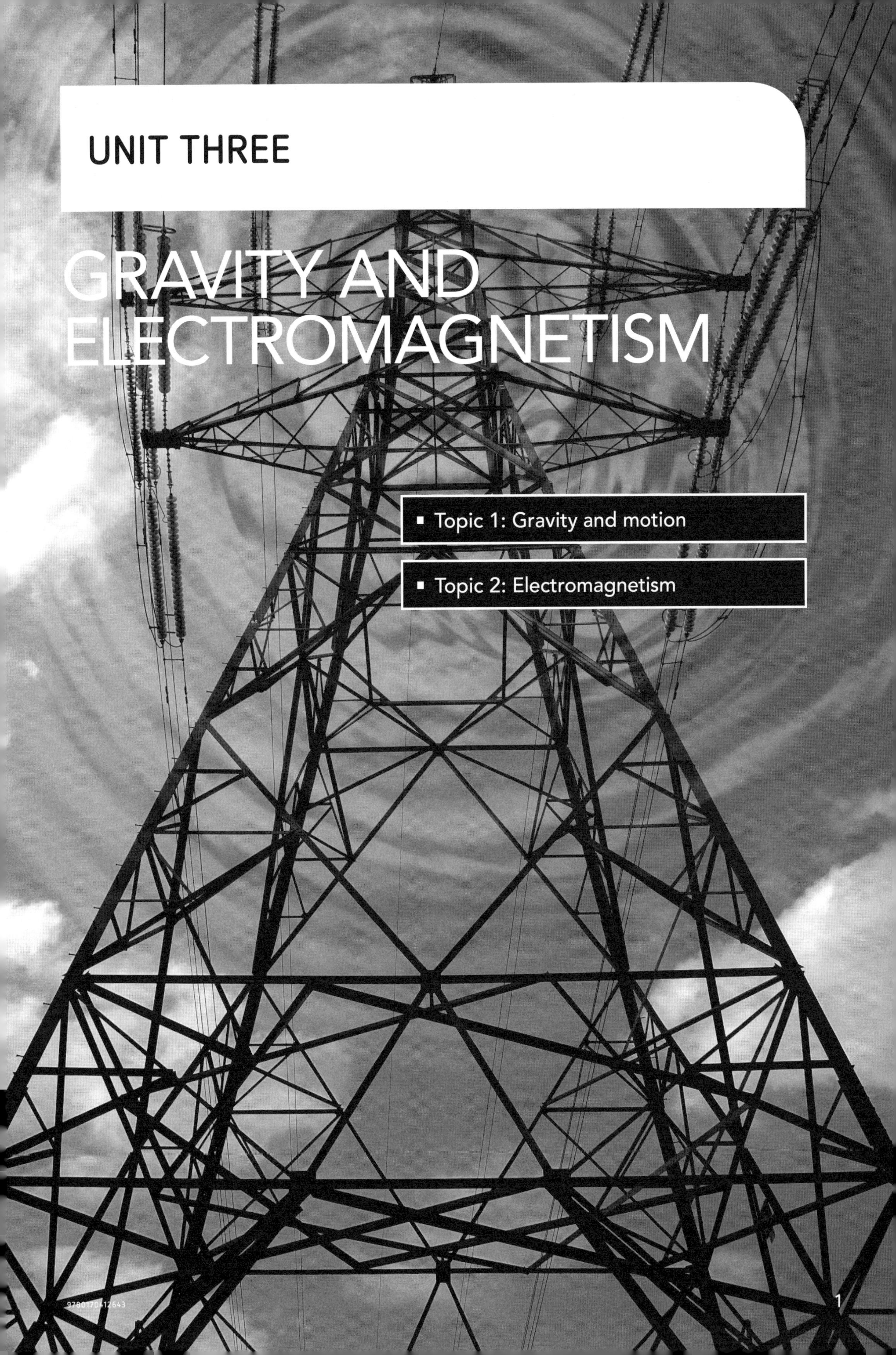

UNIT THREE

GRAVITY AND ELECTROMAGNETISM

- Topic 1: Gravity and motion
- Topic 2: Electromagnetism

1 Vector analysis

Summary

Vector analysis:

- A scalar quantity is specified by one scale.
- A vector quantity is completely specified by two or more separate scales.

Scalar and vector quantities:

- Vectors are distinguished symbolically from scalars by the arrow above the letter symbol: $\vec{A}$, $\vec{B}$, $\vec{C}$, etc.
- In kinematics and dynamics, vectors have magnitude and direction.
- In one dimension, vector quantities reduce to positive and negative numbers.
- Common vector and scalar quantities in kinematics and dynamics:

	SCALAR	VECTOR	UNIT
KINEMATICS	distance	displacement	M
	speed	velocity	$m\,s^{-1}$
	acceleration	acceleration	$m\,s^{-2}$
DYNAMICS	work–energy	(–)	N m; joule, J
	(–)	impulse–momentum	N s; $kg\,m\,s^{-1}$
	(–)	force	newton, N

Direction is applied using an agreed axis system.

- Compass points of quadrant bearings: Nθ°E, Nθ°W, Sθ°E and Sθ°W; and true bearings (azimuth): north (0° or 360°), east (90°), south (180°) and west (270°)
- Cartesian axes: positive x-axis
- Plane of a sloping surface
- One of the vectors involved

Vectors in two dimensions:

- Addition: $\vec{C} = \vec{A} + \vec{B}$
- Subtraction (inverse addition; addition of the negative):

$$\vec{A} = \vec{C} + {}^{-}\vec{B}$$
$$\Rightarrow \vec{A} = \vec{C} - \vec{B}$$

- Multiplication by a scalar: $\vec{A} \rightarrow k \times \vec{A}$

9780170412643

- The scalar length of $\vec{A}$ changes according to the magnitude of k.
- Direction changes according to the sign of k:
  - if $k > 1$ magnitude is increased but the direction is not altered
  - if $0 < k < 1$ magnitude is reduced but the direction is not altered
  - if $k < 0$ magnitude is affected *and* the direction is reversed.

Parallelogram method: Align both vectors with their tails at the same position. Construct a parallelogram using the vectors as adjacent sides. The resultant is the diagonal that starts at the tails of the vectors.

Components of vectors: Any vector can be resolved into two components or resolutes.

Rectangular components: Components at right angles to each other are called rectangular components. The geometry and trigonometry of right-angled triangles can be utilised:

1 Vector resolutes using compass bearings: resolutes are usually taken with respect to N–S and E–W.
2 Vector resolutes using the Cartesian grid: resolutes are taken with respect to the x- and y-axis respectively.
   - Rectangular components of vector, $\vec{A}$:

$$\text{magnitude of } x\text{-component: } A_x = A\cos\theta$$

$$\text{magnitude of } y\text{-component: } A_y = A\sin\theta$$

   - Magnitude of vector, $\vec{A}$: $A = \sqrt{A_x^2 + A_y^2}$
   - Angle θ, with respect to the positive direction of the x-axis.

$$\tan\theta = \frac{y\text{-component}}{x\text{-component}}$$

$$\Rightarrow \theta = \tan^{-1}\left(\frac{y\text{-component}}{x\text{-component}}\right)$$

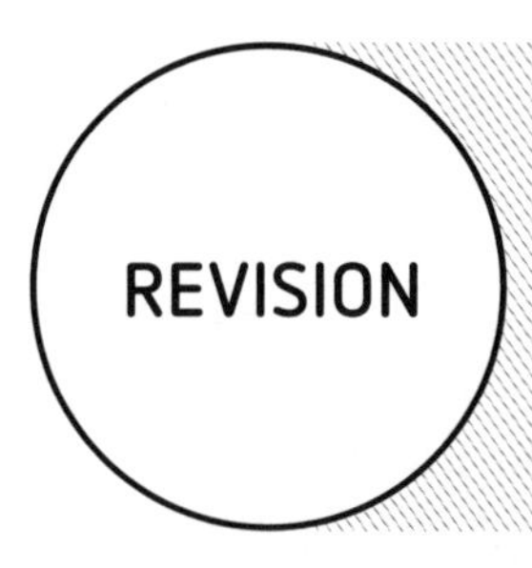

1.1 Bearings

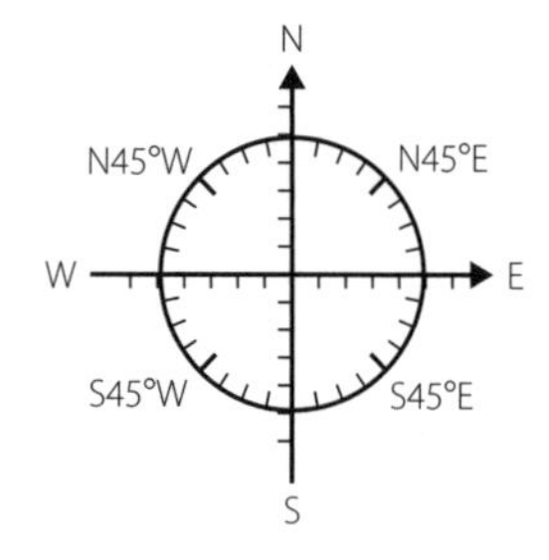

FIGURE 1.1.1 Quadrant bearings

FIGURE 1.1.2 True bearings

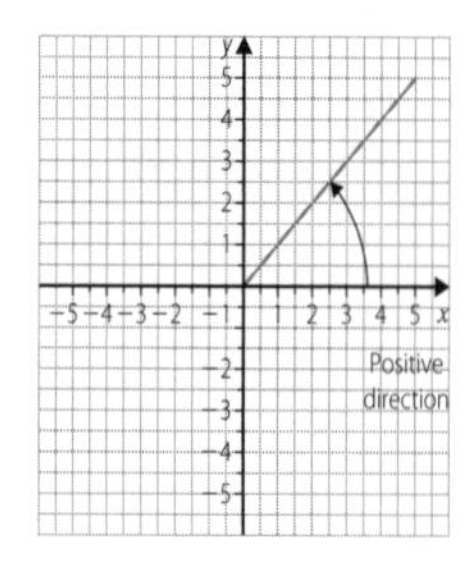

FIGURE 1.1.3 Cartesian plane

WORKED EXAMPLE

A displacement vector has a magnitude of 5.0 km and direction N30°E. Represent this vector using the following systems.

1 True bearing
2 Quadrant bearing
3 Cartesian coordinates, where the x-axis represents the direction of east

ANSWERS

1

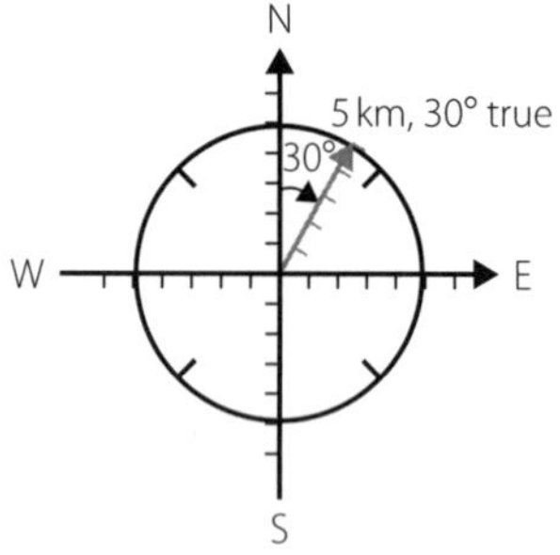

2

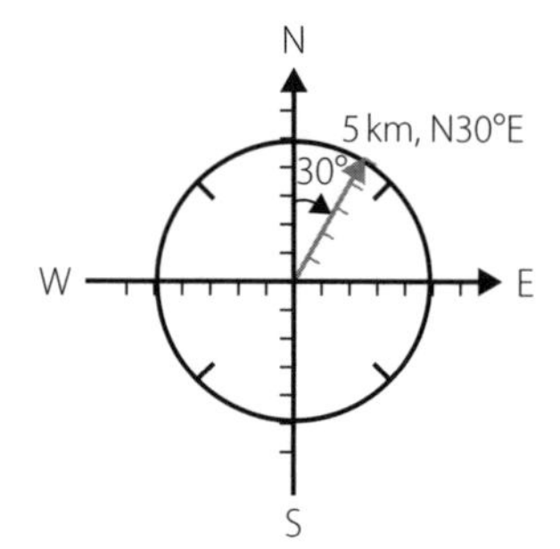

3

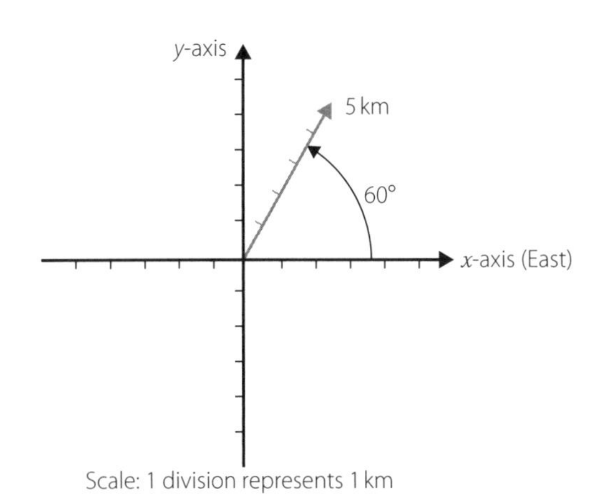

QUESTIONS

1 Draw the following vectors:

a $20\,\text{m}\,\text{s}^{-1}$, N60°E

b $20 \, m \, s^{-1}$, 250° true

2

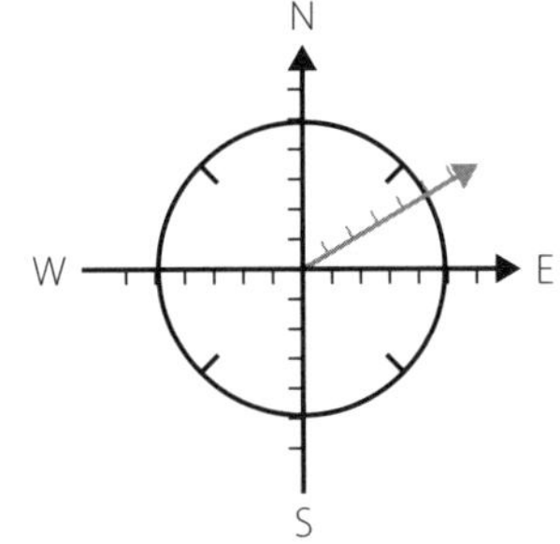

Scale: 1 unit represents 1 m

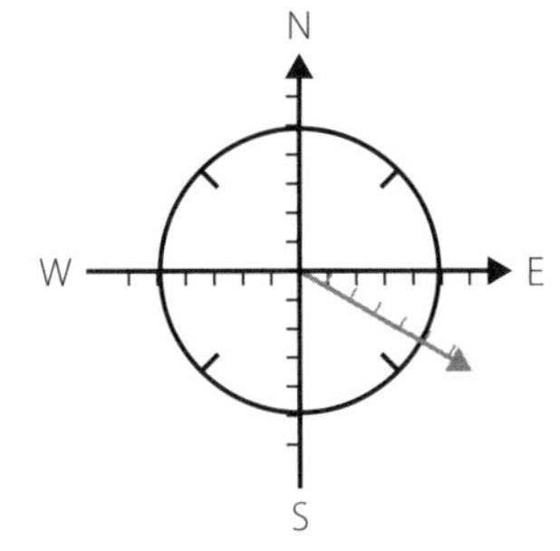

Scale: 1 unit represents 5.0 km

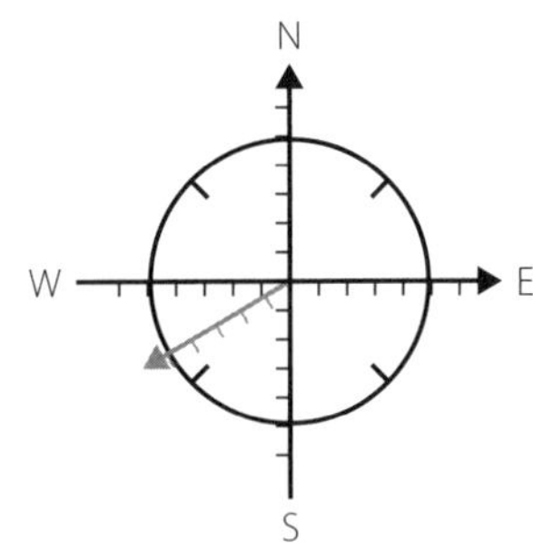

Scale: 1 unit represents 10 m

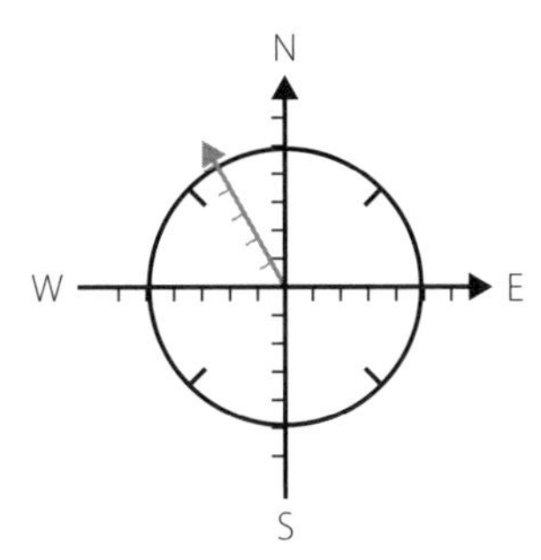

Scale: 1 unit represents 2.5 km

For each of the four vectors shown above:

a state the magnitude

____

____

b specify the direction in:

i quadrant bearing

____

____

ii true bearing.

____

____

3 Convert the following quadrant bearings to true bearings:

a N40°E

____

b N30°W

____

c S30°E

____

d S20°W.

____

4 Convert the following true bearings to quadrant bearings:

a 50° true

____

b 150° true

____

c 230° true

____

d 320° true.

____

5 The following diagram shows four straight lines of equal magnitude, P, Q, R and S, drawn on a Cartesian plane.

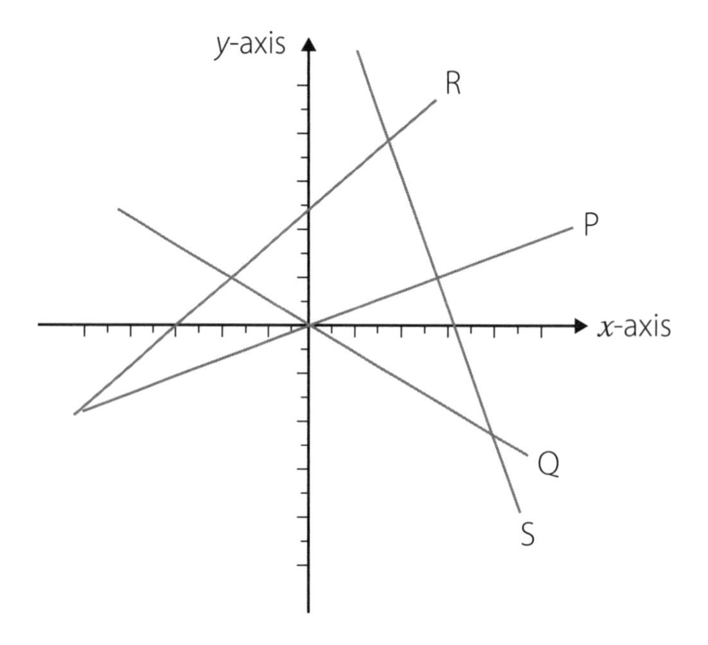

a Measure the angle for each of the lines.

____

____

b What is the range of angles that can be specified on the Cartesian plane? Use the definition of the angle to explain your reasoning.

____

____

____

9780170412643

1.2 Vectors in two dimensions

TIP-TO-TAIL METHOD

WORKED EXAMPLE

If $\vec{A} = 10\,\text{km}$, 30° true and $\vec{B} = 15\,\text{km}$, 315° true, use the tail-to-tip method to find:

1 $\vec{A} + 2\vec{B}$

2 $2\vec{A} - \vec{B}$.

ANSWERS

1

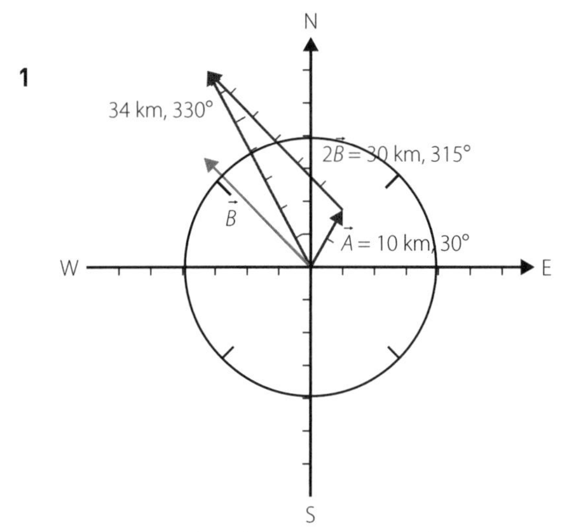

Scale: 1 division represents 5 km

Note the magnitude, but not direction, of $\vec{B}$ has been changed (×2).

2

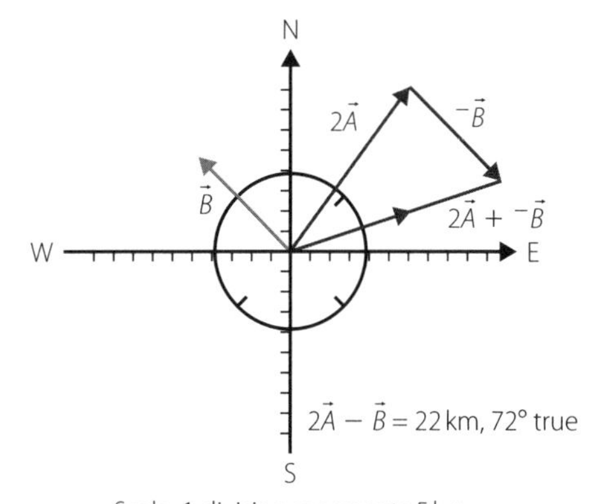

Scale: 1 division represents 5 km

Note the magnitude, but not direction of $\vec{A}$, has been changed (×2). The direction of $\vec{B}$ has been reversed but its magnitude is unchanged (x^{-1}).

PARALLELOGRAM METHOD

WORKED EXAMPLES

If $\vec{A} = 150\,\text{kg m s}^{-1}$, N30°W and $\vec{B} = 100\,\text{kg m s}^{-1}$, N60°E use the parallelogram method to find:

1 $2\vec{A} + \vec{B}$

2 $\vec{B} - 2\vec{A}$.

ANSWERS

1

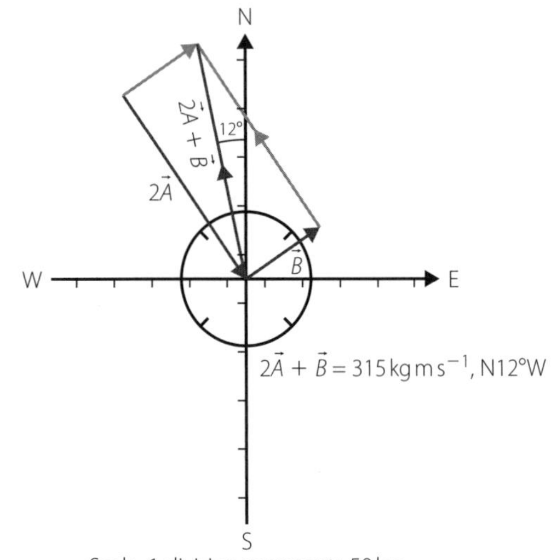

Note the magnitude, but not direction, of $\vec{A}$ has been changed (×2).

2

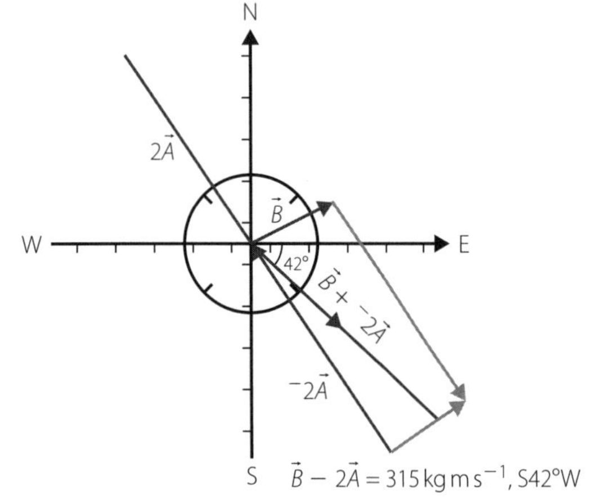

Note the magnitude of $\vec{A}$ has been changed (×2). The direction of $\vec{A}$ has been reversed, x^{-1}.

9780170412643

QUESTIONS

1 If $\vec{P} = 25\,\text{km}\,\text{h}^{-1}$, N30°E and $\vec{Q} = 25\,\text{km}\,\text{h}^{-1}$, N60°E use the tail-to-tip method to find:

a $\vec{P} + \vec{Q}$

b $3\vec{P} + 2\vec{Q}$

c $\vec{P}+4\vec{Q}$

d $\vec{P}-\vec{Q}$

9780170412643

e $\vec{Q} - \vec{P}$

f $2\vec{P} - 3\vec{Q}$.

2 If $\vec{S} = 6.0\,\text{kg}\,\text{m}\,\text{s}^{-1}$, 45° true and $\vec{T} = 8.0\,\text{kg}\,\text{m}\,\text{s}^{-1}$, 315° true, use the parallelogram method to find:

a $\vec{S} + \vec{T}$

b $2\vec{S} + \vec{T}$

9780170412643

c $3\vec{S}+4\vec{T}$

d $\vec{S}-\vec{T}$

e $3\vec{S} - 2\vec{T}$

f $2\vec{T} - 4\vec{S}$.

9780170412643

3 If $\vec{G} = 1.6 \times 10^3$ N, 45° true and $\vec{H} = 8.0 \times 10^3$ N, 135° true use the parallelogram method to find:

a $\vec{G} - \vec{H}$

b $\vec{G} - 2\vec{H}$

c $4\vec{G}+3\vec{H}$

d $\vec{G}+0.5\vec{H}$

9780170412643

e $0.8\vec{G}+2.5\vec{H}$

f $6\vec{G}+11\vec{H}$.

1.3 Components of vectors

WORKED EXAMPLES

1 For the vectors below, find appropriate rectangular resolutes by scale drawing.

a 50 N, 35° true

b 15 $kg\,m\,s^{-1}$, N55°W

c 80 m, 120°

2 Two vectors are shown on a Cartesian plane. Show how to add the two vectors using components.

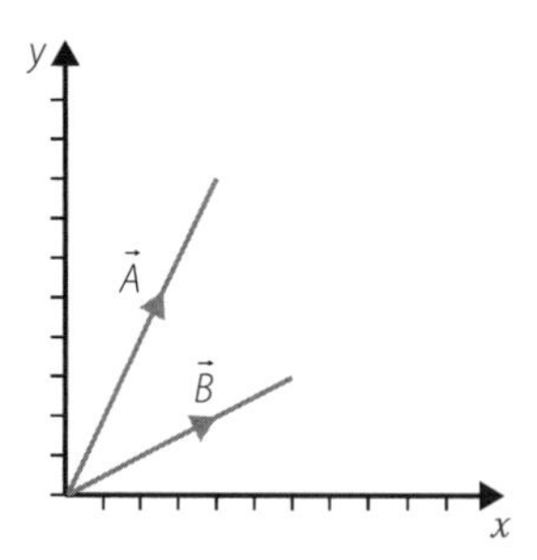

ANSWERS

1 a

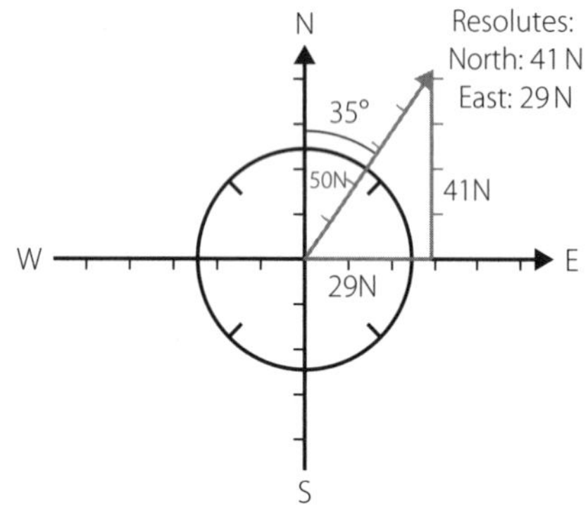

Scale: 1 division represents 10 N

b

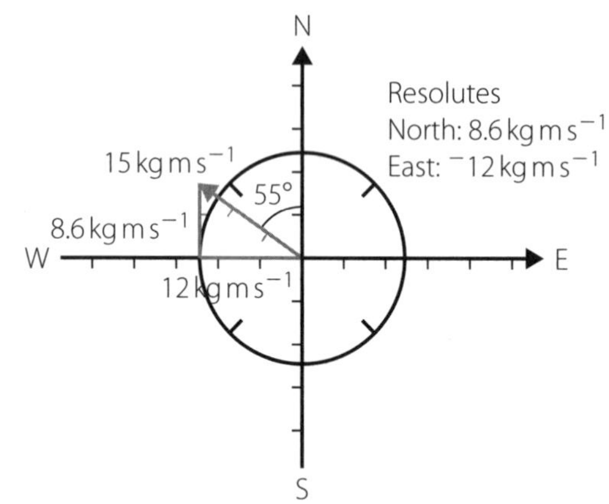

Scale: 1 division represents 5 $kg\,m\,s^{-1}$

c

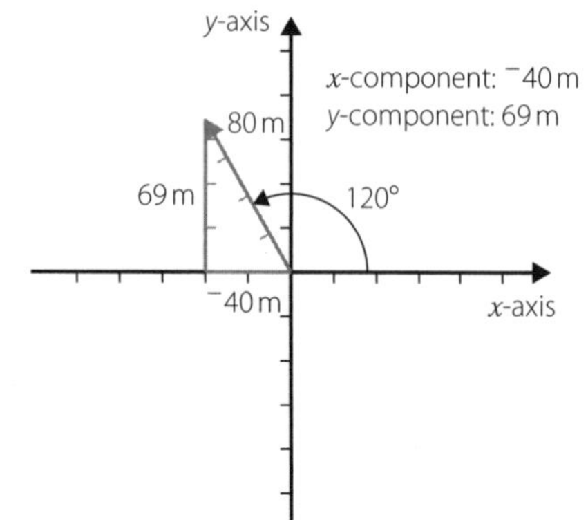

Scale: 1 division represents 20 m

2

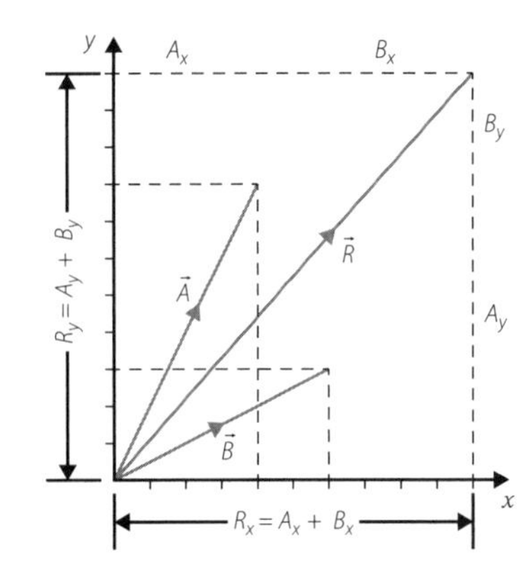

QUESTIONS

1 On a Cartesian plane, $\vec{A} = 6.0$ N, 30° and $\vec{B} = 8.0$ N, 60°. Draw each vector onto a Cartesian plane and show their rectangular components. Find the resultant vector by scale measurement.

a $\vec{A} + \vec{B}$

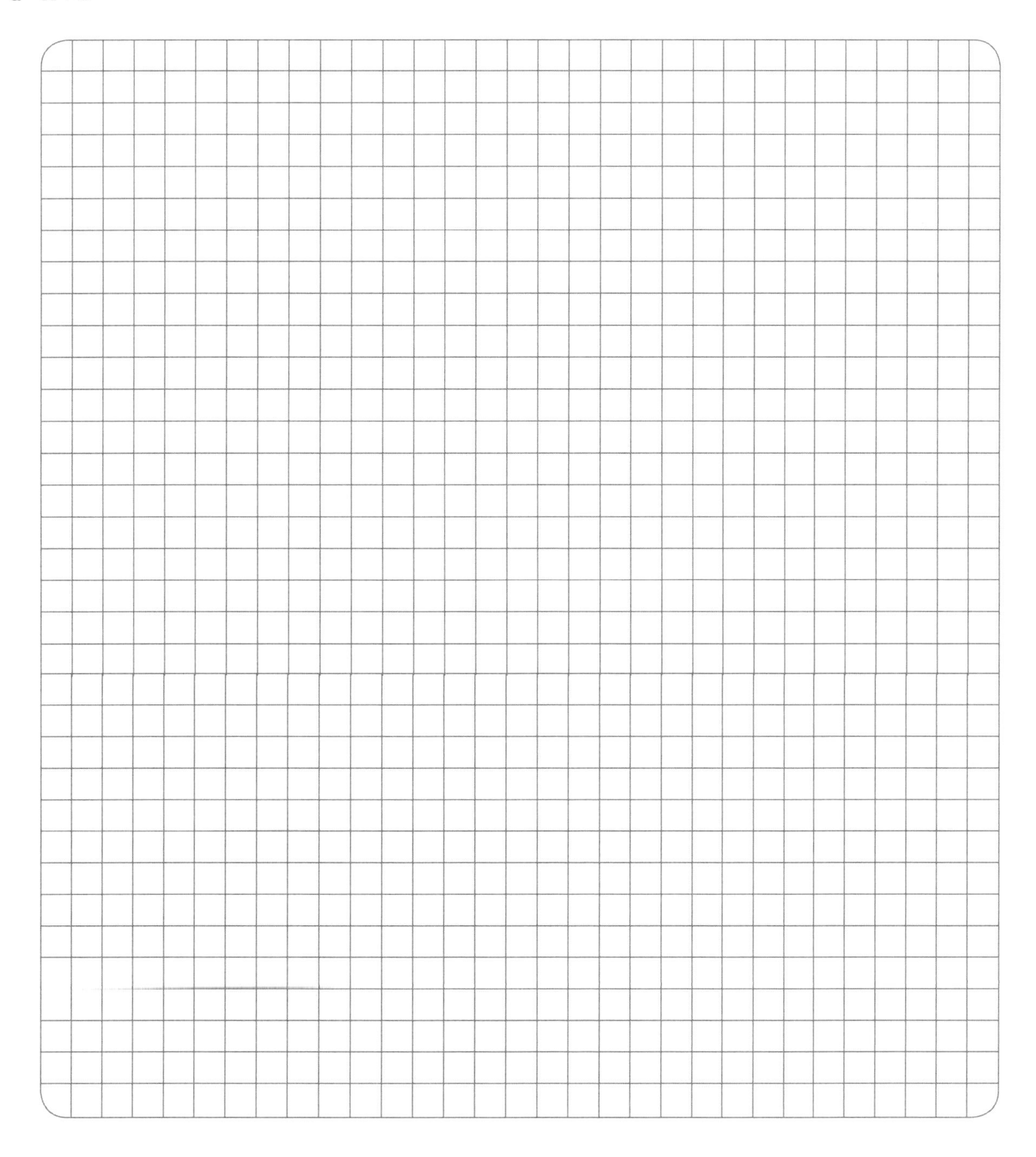

b $\vec{A}-\vec{B}$

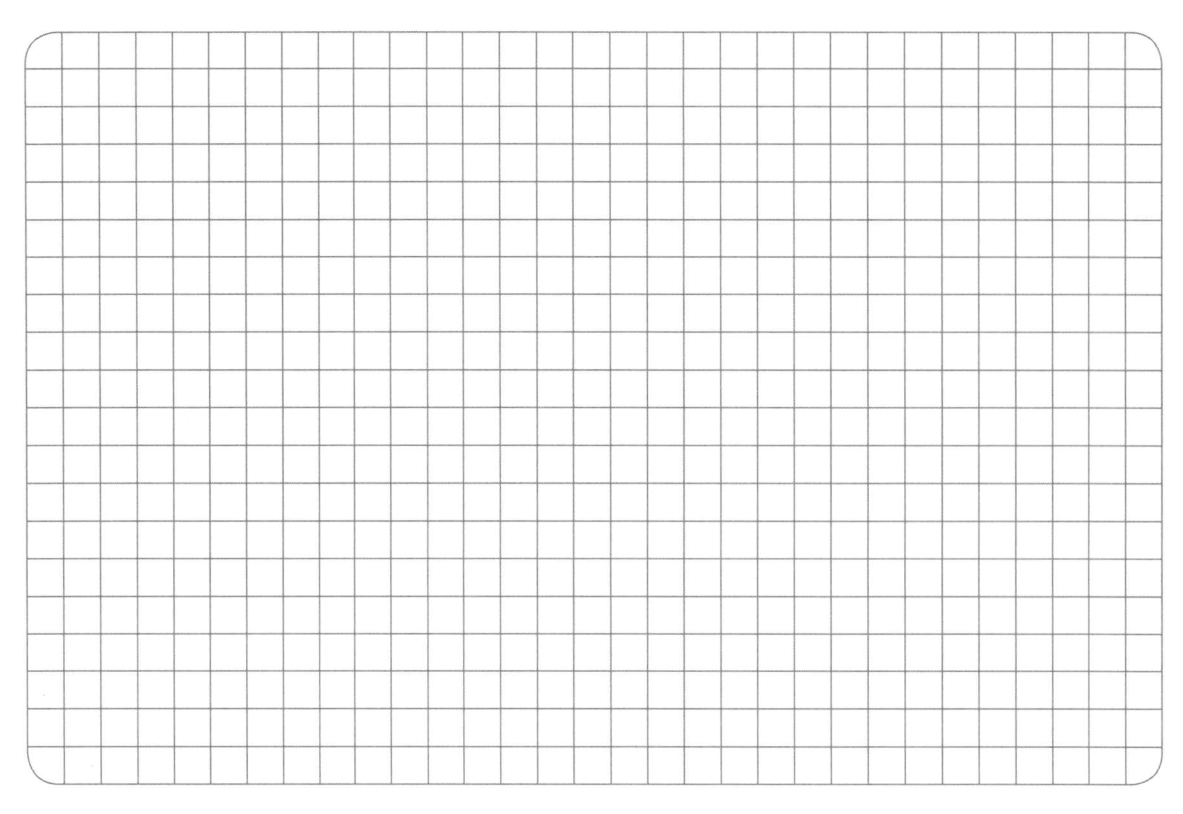

2 If $\vec{P}=2.0\times10^{-4}\,\mathrm{kg\,m\,s^{-1}}$, 135° and $\vec{Q}=4.0\times10^{-4}\,\mathrm{kg\,m\,s^{-1}}$, 30° use the parallelogram method on a Cartesian plane to find:

a $\vec{P}+\vec{Q}$

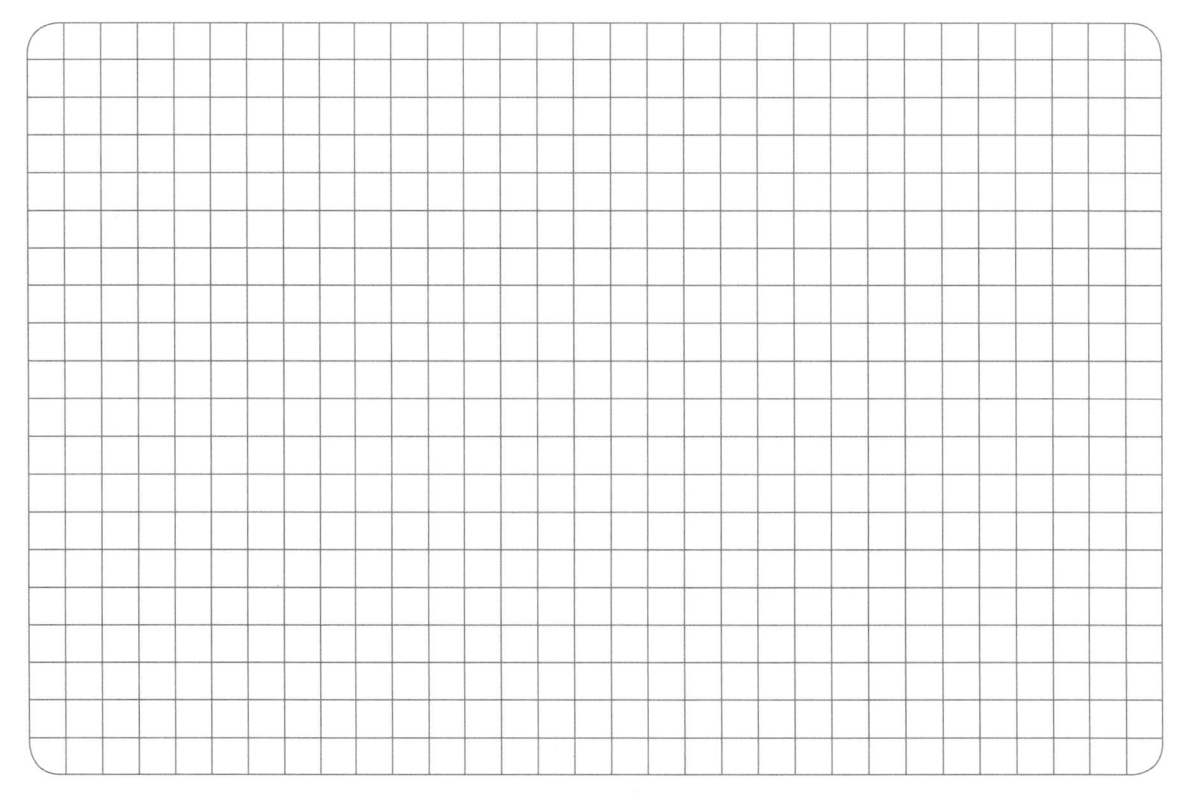

b $\vec{P} - \vec{Q}$

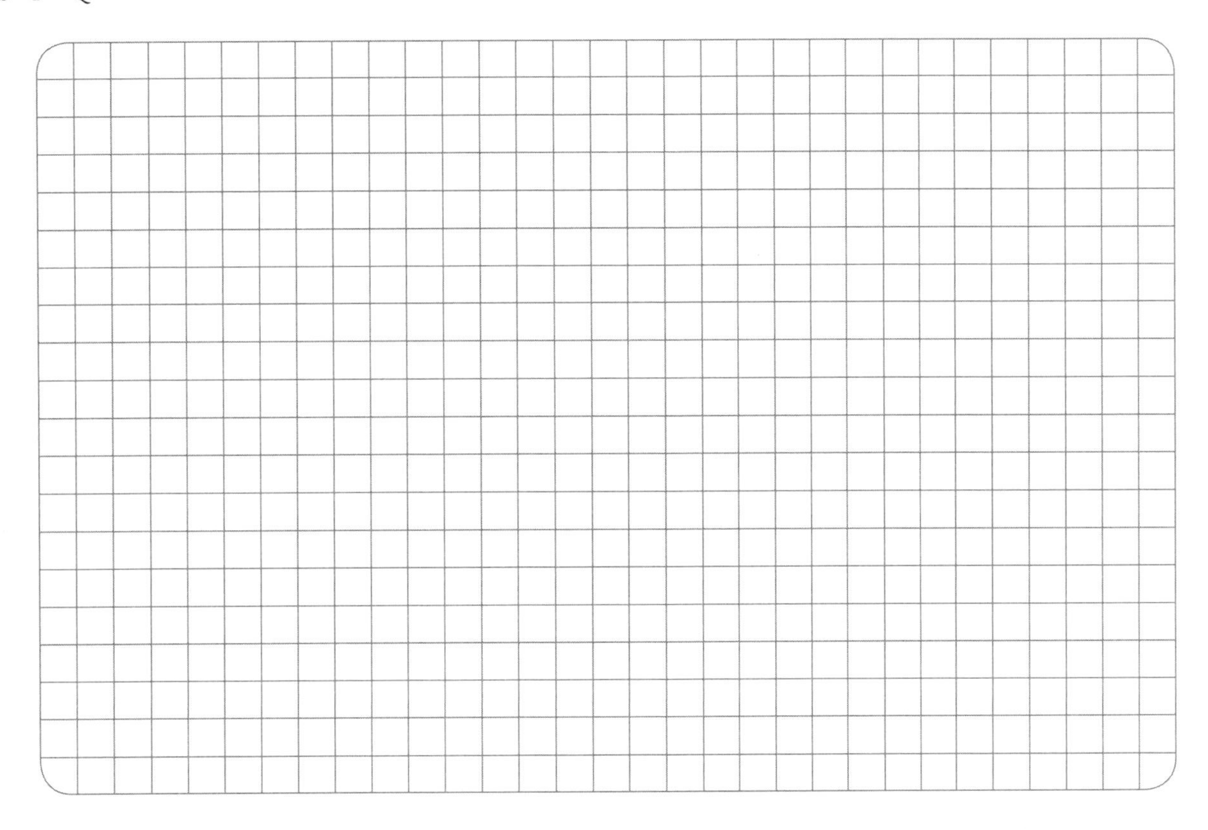

3 Use rectangular components to add the vectors shown on the Cartesian plane.

a

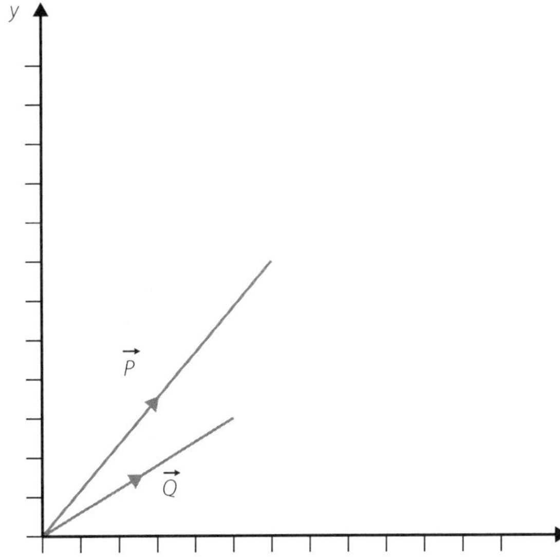

b

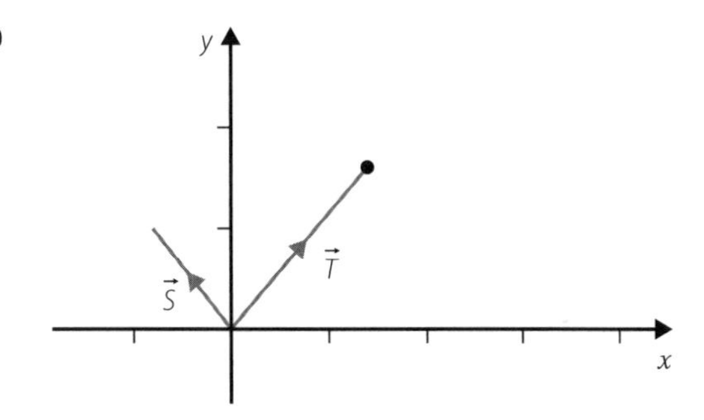

4 Two vectors, $\vec{A}$ and $\vec{B}$ have components $A_x = 56\,\text{N}$; $A_y = 72\,\text{N}$ and $B_x = 56\,\text{N}$; $B_y = 72\,\text{N}$ respectively. Calculate the following:

a $\vec{A} + \vec{B}$

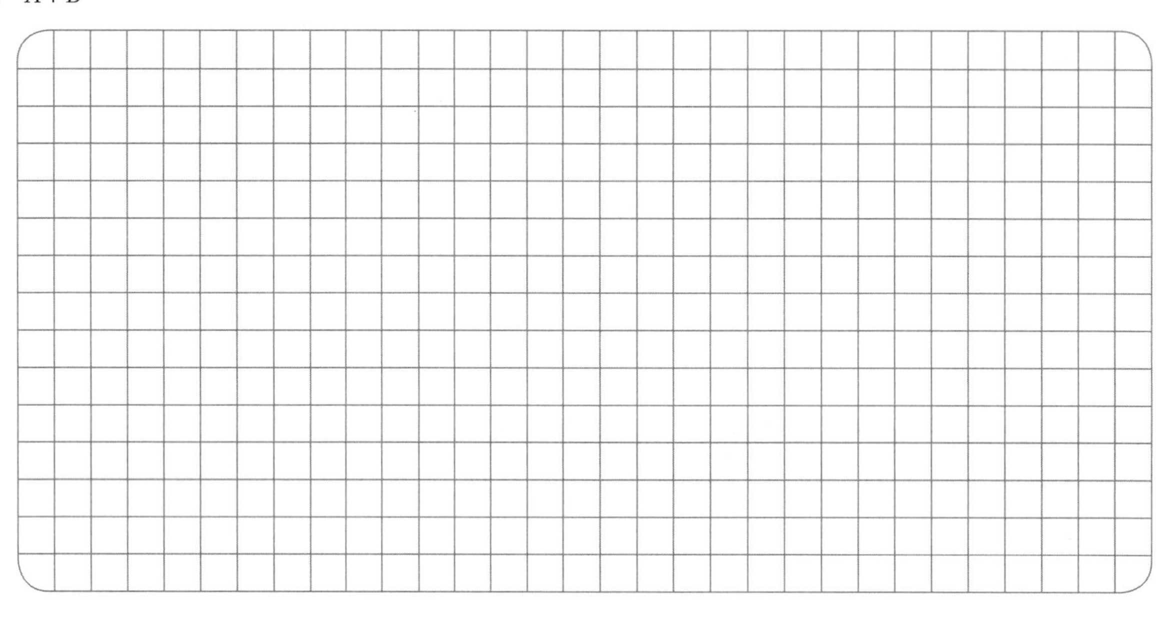

b $\vec{A} - \vec{B}$

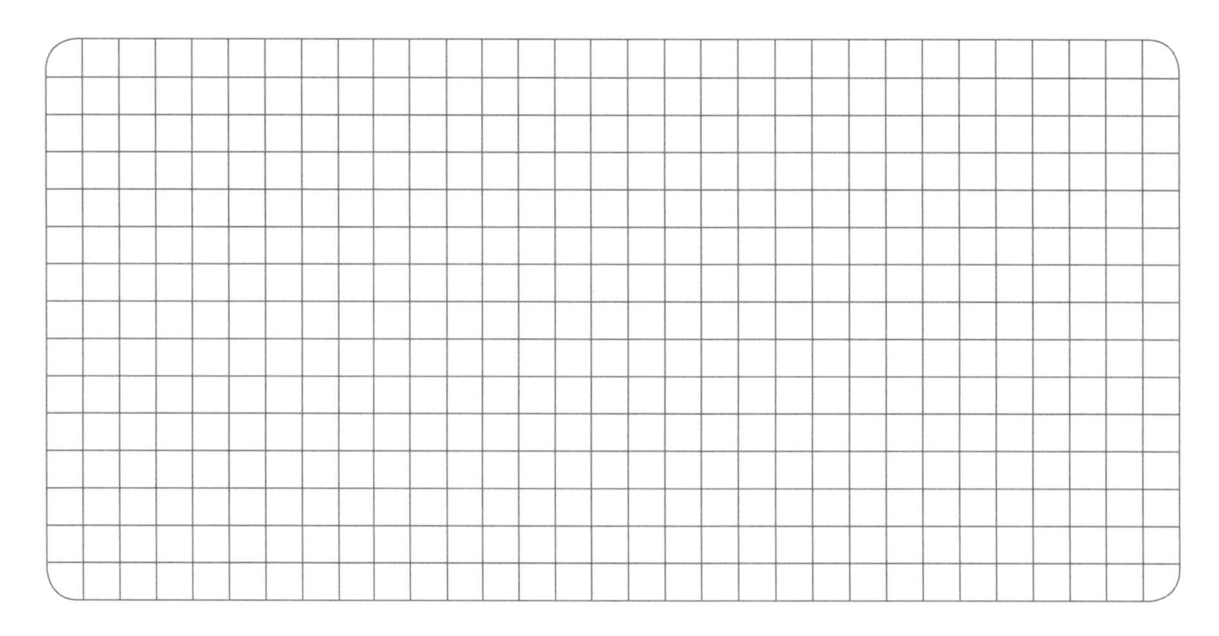

c $\vec{B}+2\vec{A}$

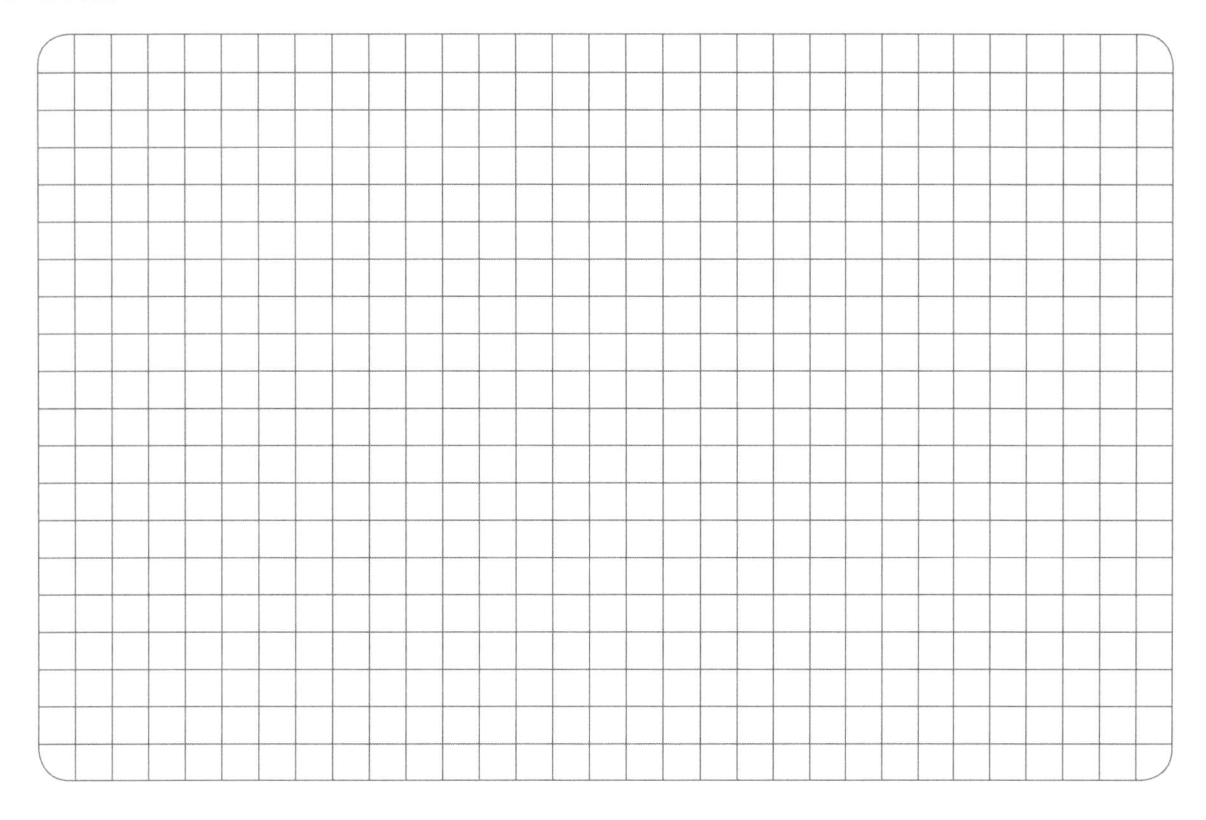

d $\vec{A}-2\vec{B}$.

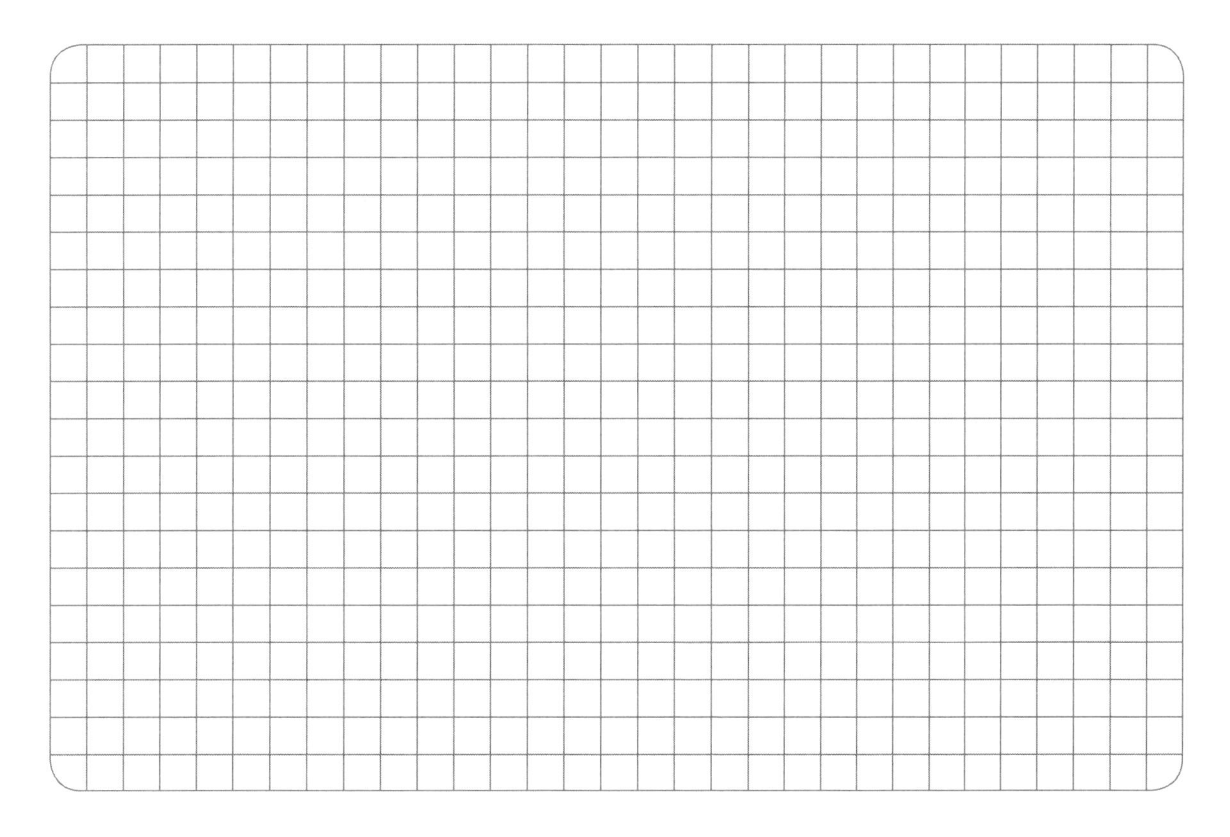

5 If $\vec{P} = 3.0 \times 10^3 \text{ kg m s}^{-1}$, 100° and $\vec{Q} = 4.0 \times 10^3 \text{ kg m s}^{-1}$, $^{-}60°$ find the components of:

a $\vec{P}$

b $\vec{Q}$.

c Hence, find the following vector additions and subtractions:

i $\vec{P} + \vec{Q}$

ii $\vec{P} - \vec{Q}$

iii $2\vec{P} - \vec{Q}$

iv $0.5\vec{Q} - \vec{P}$

Multiple-choice

1 The angle associated with the vector shown in the figure below is:

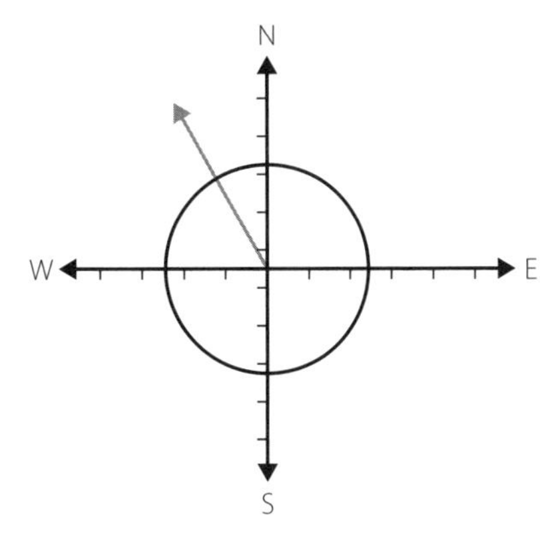

A N30°W.

B W30°N.

C 30° true.

D 120° true.

2 In the figure below the resolute of the vector in the easterly direction is:

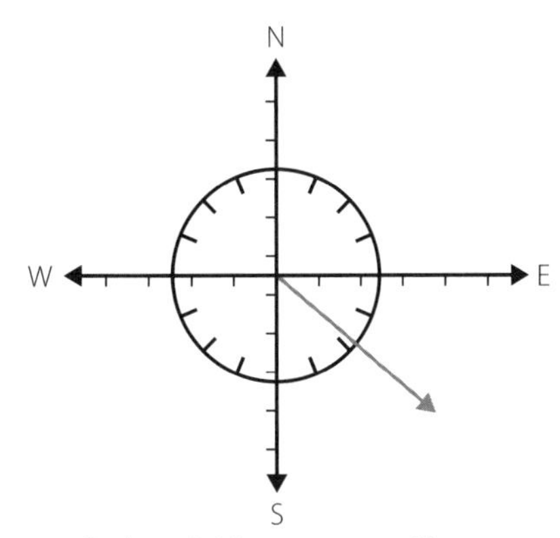

Scale: 1 division represents 1 km

A $^{+}3.8$ km.

B $^{-}3.8$ km.

C $^{+}3.2$ km.

D $^{-}3.2$ km.

3 The figure below shows a small mass, which is free to slide, on a plane that is inclined at 30° to the horizontal. The resolute of the acceleration due to gravity parallel to the plane is:

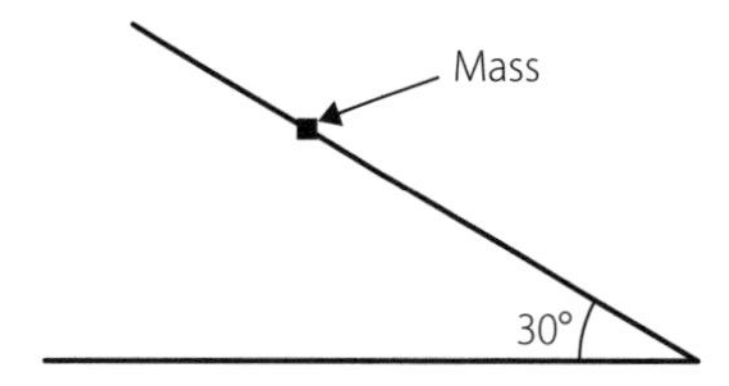

A $9.8\,\text{m}\,\text{s}^{-2}$.

B $9.8 \times \sin 30°\,\text{m}\,\text{s}^{-2}$.

C $9.8 \times \cos 30°\,\text{m}\,\text{s}^{-2}$.

D $9.8 \times \tan 30°\,\text{m}\,\text{s}^{-2}$.

4 A mass slides freely on a plane that is inclined at 60° to the horizontal. The acceleration due to gravity is:

A $4.9\,\text{m}\,\text{s}^{-2}$.

B $8.5\,\text{m}\,\text{s}^{-2}$.

C $9.8\,\text{m}\,\text{s}^{-2}$.

D $17\,\text{m}\,\text{s}^{-2}$.

5 The figure below shows a vector on a Cartesian plane. The angle associated with this vector is:

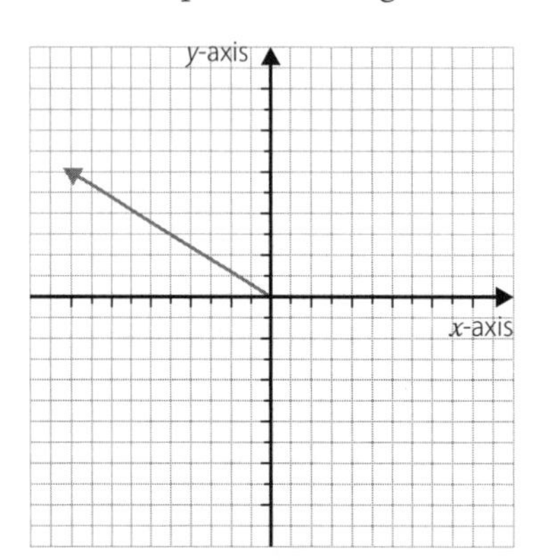

A 30°.

B 60°.

C 120°.

D 150°.

6 The figure below shows two vectors, $\vec{A}$ and $\vec{B}$.

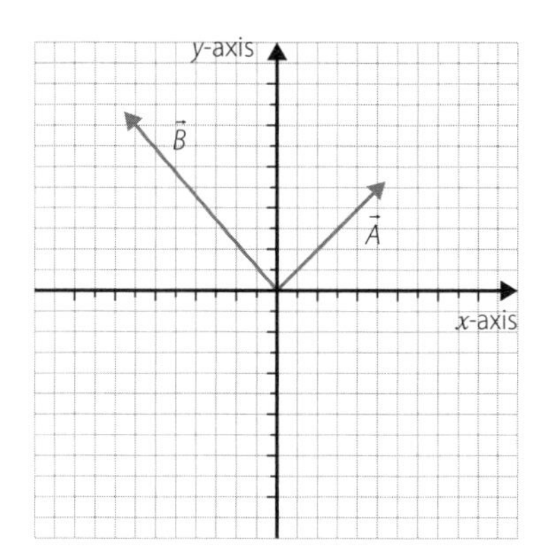

The x component of the resultant vector, $\vec{R} = \vec{B} - 2\vec{A}$ is:

A $^{-}18$

B $^{-}13$

C $^{-}3$

D 2

7 Vectors $\vec{P}$ and $\vec{Q}$ are drawn on a Cartesian plane. Vector $\vec{P} = 5.0$ units at 30° and vector $\vec{Q} = 8.0$ units at 120° When added, the x-component of the sum is:

A $^{+}8.3$ units.

B $^{+}0.3$ units.

C $^{-}8.3$ units.

D $^{-}0.3$ units.

8 A force of 10 N along the x-axis is added to a force of 20 N applied along the y-axis. The sum of these forces is:

A 30 N, 45°.

B 30 N, 63°.

C 22 N, 45°.

D 22 N, 63°.

9 The sum of two vectors, $\vec{S}$ and $\vec{T}$, is 80 kg m s^{-1}, 90° true. If $\vec{S} = 80$ m s^{-1}, 210° true, vector $\vec{T}$ =?

A 120 kg m s^{-1}, 30° true

B 140 kg m s^{-1}, 30° true

C 120 kg m s^{-1}, 60° true

D 140 kg m s^{-1}, 60° true

Short answer

10 A mass is subjected to two forces, $\vec{P}$ and $\vec{Q}$, where $\vec{P} = 100\,\text{N}$, S30°W and $\vec{Q} = 200\,\text{N}$, 120° true.

a Specify the components of both $\vec{P}$ and $\vec{Q}$ with respect to the positive directions of north and east, using a geometric method.

b Find the resultant, $\vec{R} = \vec{P} + \vec{Q}$, using a geometric method.

c Find the resultant, $\vec{R} = 2\vec{P} + \vec{Q}$, using the components method.

11 One force of 10 N is applied at an angle of 45° to a second force of 20 N. Use a scale drawing to show:

a the magnitude and direction of the combined force, relative to the 10 N force

b the magnitude of the component of the 10 N force, which is:

i parallel to the 20 N force

ii perpendicular to the 20 N force.

c On a scale drawing show the sum of the components of the 10 N force and the 20 N force, which are parallel to the resultant.

12 A 400 g mass collides with a stationary 600 g mass at $12.5\,\mathrm{m\,s^{-1}}$. After the collision, the 400 g mass has a momentum of $2.0\,\mathrm{kg\,m\,s^{-1}}$ at an angle of 30°. The 600 g mass moves off at momentum, $\vec{p}_f$(magnitude, p_f; angle, θ). The sum of the momenta of the two masses after the collision is $5.0\,\mathrm{kg\,m\,s^{-1}}$ at 0°. All angles are measured relative to the original path of the 400 g mass.

a Sketch and annotate a diagram to represent the situation described.

b For the situation *after* the collision, specify:

i the component of the momentum of the 400 g mass along the line of its original motion

______

______

______

ii the component of the momentum of the 600 g mass along the line of the original motion of the 400 g mass.

______

______

______

13 A $1.2 \times 10^{2}\,\mathrm{kg}$ car accelerates down a 50° slope at $2.5\,\mathrm{m\,s^{-2}}$. The acceleration is caused by the combined effect of three forces: weight, the force normal to the surface and friction, which acts parallel to the surface. For this question, use $g = 10\,\mathrm{m\,s^{-2}}$.

a In the space provided below, draw and annotate one or more scale drawings in order to specify completely the following:

i weight

ii the component of the weight perpendicular to the surface

iii the component of the weight parallel to the surface

iv friction

v net force.

b What quantitative effect does the friction have on the acceleration of the car down the slope?

2 Projectile motion

Summary

- Projectile motion occurs on a vertical plane, 'near Earth'.
- Near Earth the acceleration due to gravity is constant: $g = 9.8\,\text{m}\,\text{s}^{-2}$.
- Cartesian coordinates, (x, y), are used to identify the position of a projectile.
  - x is for horizontal direction
  - y is for vertical direction
- Projectile motion can be resolved into two rectangular components:
  - the horizontal component is *not* affected by gravity
  - the vertical component is only affected by gravity.
- The horizontal (x) and vertical (y) components are independent of each other.
- For a projectile launched at speed, u, and angle, θ, relative to the horizontal:
  - the horizontal component of the launch velocity is $u_x = u\cos\theta$
  - the vertical component the launch velocity is $u_y = u\sin\theta$.
- When combining vertical and horizontal components of motion, at any position (x, y) along its path, the velocity, $\vec{v}_{(x,y)}$, of the projectile is given by:
  - magnitude: $v_{(x,y)} = \sqrt{v_x^2 + v_y^2}$
  - direction: $\theta = \tan^{-1}\left(\frac{v_y}{v_x}\right)$
  - where: $v_{(x,y)}$ = speed at position (x, y); v_x = x-component of velocity at position (x, y); v_y = y-component of velocity at position (x, y); θ = angle relative to horizontal.

REVISION

2.1 Motion in the horizontal direction

WORKED EXAMPLES

1 For a projectile that is launched at an angle of elevation of 30° and speed 20.0 m s^{-1}, find the horizontal component of the launch velocity.

2 Find the launch speed for a projectile that takes 4.0 s to travel 52 m when projected at 60° to the horizontal.

ANSWERS

1 $u_x = u\cos\theta$

$u_x = 20.0\,\text{m}\,\text{s}^{-1} \times \cos 30°$

$u_x = 17.3\,\text{m}\,\text{s}^{-1}$

2 $u_x = \dfrac{52\,\text{m}}{4.0\,\text{s}}$

$\Rightarrow u_x = 13\,\text{m}\,\text{s}^{-1}$

$u_x = u\cos\theta$

$\Rightarrow u = \dfrac{u_x}{\cos\theta}$

$\Rightarrow u = \dfrac{13\,\text{m}\,\text{s}^{-1}}{\cos 60°}$

$\Rightarrow u = 26\,\text{m}\,\text{s}^{-1}$

QUESTIONS

1 Complete the following table.

INITIAL LAUNCH SPEED (m s^{-1})	LAUNCH ANGLE (°)	HORIZONTAL COMPONENT OF INITIAL LAUNCH SPEED (m s^{-1})
10	30	
2.5	60	
15	40	
	45	28
	70	5.6
72		24
20		4.8

2 Find the launch speed for a projectile that takes 16 s to travel a horizontal distance of 450 m when projected at 30° to the horizontal.

____

____

____

3 A ball launched at 76° takes 2.3 s to reach the top of its flight before landing 15 m away. What was the launch speed?

4 A cricket ball struck from the pitch at a speed of $32\,\text{m s}^{-1}$ lands 55 m away in 3.5 s. At what angle did it leave the pitch?

2.2 Motion in the vertical direction

WORKED EXAMPLES

1 For a projectile that is launched at an angle of elevation of 30° and speed $20.0\,\text{m s}^{-1}$ find the vertical component of the launch velocity.

2 A rock is hurled horizontally from a 50 m high cliff. For the launch, state the vertical component of the following quantities. In each case, give a reason for your answer.

a The displacement from the cliff

b The launch velocity

c The acceleration

ANSWERS

1 $u_y = u\sin\theta$

$u_y = 20.0\,\text{m s}^{-1} \times \sin 30°$

$u_y = 10.0\,\text{m s}^{-1}$

2 **a** 0 m: relative to the cliff the zero of displacement is at the cliff (positive direction is downwards).

b $0\,\text{m s}^{-1}$: the rock is about to begin its vertical descent, but has yet to start moving down.

c $a = {}^{+}9.8\,\text{m s}^{-2}$: increasing downward displacement corresponds with an increasing vertical component of motion, hence, corresponds with a positive downward acceleration (there is no horizontal component of the acceleration).

QUESTIONS

1 Complete the table below.

INITIAL LAUNCH SPEED (m s^{-1})	LAUNCH ANGLE (°)	VERTICAL COMPONENT OF INITIAL LAUNCH SPEED (m s^{-1})
10	30	
2.5	60	
15	40	
	45	28
	70	5.6
72		24
20		4.8

2 A ball launched from level ground at 70° takes 3.91 s to rise to a maximum height of 75 m. What was the ball's launch speed?

____

____

____

____

____

____

2.3 Algebraic analysis of projectile motion

WORKED EXAMPLES

1 Find the velocity of a projectile after 2.0 s of flight, given that the projectile was launched at a velocity of $30\,\text{m s}^{-1}$ and angle of elevation 60°.

2 How far does a ball travel over level ground when launched at speed $15.0\,\text{m s}^{-1}$ and angle 30° from the ground?

ANSWERS

1 Horizontally

$$u_x = u\cos\theta$$

$$u_x = 30\,\text{ms}^{-1} \times \cos 60°$$

$$u_x = 15\,\text{ms}^{-1}$$

At all times, this is the horizontal component of the projectile.

Vertically

$s = ?\ u_y = u\sin\theta\ v_y = ?\ a = {}^{-}9.8\,\text{ms}^{-2}\ t = 2.0\,\text{s}$

$u_y = 30\,\text{ms}^{-1} \times \sin 60°$

$u_y = 26\,\text{ms}^{-1}$

$v_y = gt + u_y$

$\Rightarrow v_y = {}^{-}9.8\,\text{ms}^{-2} \times 2.0\,\text{s} + 26\,\text{ms}^{-1}$

$\Rightarrow v_y = 6.4\,\text{ms}^{-1}$

Combining horizontal and vertical components

$\Rightarrow v_{(x,y)} = \sqrt{v_x^2 + v_y^2}$

$\Rightarrow v_{(x,y)} = \sqrt{(15\,\text{ms}^{-1})^2 + (6.4\,\text{ms}^{-1})^2}$

$\Rightarrow v_{(x,y)} = 16.3\,\text{ms}^{-1}$

$\theta = \tan^{-1}\left(\frac{v_y}{v_x}\right)$

$\Rightarrow \theta = \tan^{-1}\left(\frac{6.4\,\text{ms}^{-1}}{15\,\text{ms}^{-1} v_x}\right)$

$\Rightarrow \theta = 23°$

2 $R = \frac{u^2 \sin 2\theta}{g}$

$\Rightarrow R = \frac{(15.0\,\text{ms}^{-1})^2 \times \sin(2 \times 30°)}{9.8\,\text{ms}^{-2}}$

$\Rightarrow R = 11.5\,\text{m}$

QUESTIONS

1 Complete the table below.

INITIAL LAUNCH SPEED (m s^{-1})	LAUNCH ANGLE (°)	HORIZONTAL COMPONENT OF INITIAL LAUNCH SPEED (m s^{-1})	VERTICAL COMPONENT OF INITIAL LAUNCH SPEED (m s^{-1})
40	60		
	30	10	
28			18
		15	12
		2.7	4.3

2 A 6.8 g stone is thrown horizontally at $21.0\,\text{m s}^{-1}$ from a cliff, which is 28.0 m high. Calculate the:

a time taken to reach the ground

9780170412643

b horizontal distance travelled before reaching the ground

c velocity after 14.0 m of travel.

3 A rocket is propelled upwards towards a cliff at 65.0 $m s^{-1}$ and angle 60°. After reaching a maximum height it strikes the cliff 45 m above the launch position. Find the:

a maximum height attained

b time to reach maximum height

c distance to the cliff

d time taken to strike the cliff.

4 A person standing 25.0 m above a lake throws a stone downwards at an angle of 30° to the horizontal. After 1.42 s, the stone splashes into the lake, 18.0 m horizontally away from the person. Find the speed of the stone when the stone:

a leaves the person's hand

b enters the water.

5 Two identical buildings are separated by a distance of 14.5 m. A ball is thrown at an angle of 45° and speed 12.7 m s^{-1}. If the thrower is 2.0 m from the edge of the building, decide if the ball lands on the top of the other building.

____

____

____

____

2.4 Solution strategy: projectile motion

WORKED EXAMPLES

1 A small rocket is launched at a speed of 80 m s^{-1} at 70° to the horizontal. When the rocket is 200 m above the ground for the first time, find the:

a vertical component of the velocity

b speed

c direction of motion

d time of flight until it is next at 200 m above the ground.

2 A projectile originally launched at angle θ over level ground at 40.0 m s^{-1} lands 80 m away. Calculate the angle of launch.

ANSWERS

1 **a** $u_y = u\sin\theta$

$\Rightarrow u_y = 80.0\,\text{ms}^{-1} \times \sin 70°$

$\Rightarrow u_y = 75.2\,\text{ms}^{-1}$

$v_y^2 = 2as_y + u_y^2$

$\Rightarrow v_y = \sqrt{2as_y + u_y^2}$

$\Rightarrow v_y = \sqrt{2 \times {}^-9.8\,\text{ms}^{-2} \times 200\,\text{m} + (75.2\,\text{ms}^{-1})^2}$

$\Rightarrow v_y = 41.6\,\text{ms}^{-1}$

b $v_{(x,y)} = \sqrt{u_x^2 + v_y^2}$

$u_x = u\cos\theta°$

$\Rightarrow u_x = 80.0\,\text{ms}^{-1} \times \cos 70°$

$\Rightarrow u_x = 27.4\,\text{ms}^{-1}$ and $u_y = 72.5\,\text{ms}^{-1}$

$\Rightarrow v_{(x,y)} = \sqrt{(27.4\,\text{ms}^{-1})^2 + (41.6\,\text{ms}^{-1})^2}$

$\Rightarrow v_{(x,y)} = 49.8\,\text{ms}^{-1}$

c $\theta = \tan^{-1}\left(\frac{v_y}{u_x}\right)$

$\Rightarrow \theta = \tan^{-1}\left(\frac{41.6}{27.4}\right)$

$\Rightarrow \theta = 56°$

d $s_y = 200\,\text{m}$ $u_y = 75.2\,\text{ms}^{-1}$ $v_y = 41.6\,\text{ms}^{-1}$ $a = {}^{-}9.8\,\text{ms}^{-2}$ $t = ?$

$$s_y = u_y t + \frac{1}{2}at^2$$

$$200\,\text{m} = (75.2\,\text{ms}^{-1}) \times t + \frac{1}{2}({}^{-}9.8\,\text{ms}^{-2}) \times t^2$$

$$4.9t^2 - 75.2t + 200 = 0$$

$$t^2 - 15.3t + 40.8 = 0$$

$$t = \frac{15.3 \pm \sqrt{(15.3)^2 - 4 \times 1 \times 40.8}}{2 \times 1}$$

$$t = 7.67 \pm 4.25$$

$$t = 3.4\,\text{s or } t = 11.9\,\text{s}$$

$$\Delta t_{200\,\text{m}} = 11.9\,\text{s} - 3.4\,\text{s}$$

$$\Delta t_{200\,\text{m}} = 8.5\,\text{s}$$

2 $R = \dfrac{u^2 \sin 2\theta}{g}$

$$\Rightarrow \sin 2\theta = \frac{gR}{u^2}$$

$$\Rightarrow 2\theta = \sin^{-1}\left(\frac{gR}{u^2}\right)$$

$$\Rightarrow \theta = \frac{1}{2}\sin^{-1}\left(\frac{gR}{u^2}\right)$$

$$\Rightarrow \theta = \frac{1}{2} \times \sin^{-1}\left(\frac{9.8\,\text{ms}^{-2} \times 80\,\text{m}}{\left(40\,\text{ms}^{-1}\right)^2}\right)$$

$$\Rightarrow \theta = 15^\circ$$

QUESTIONS

1 A stone is thrown horizontally at $9.0\,\text{m}\,\text{s}^{-1}$ from the top of a 15 m high cliff. With what velocity does it reach the bottom of the cliff?

2 The diagram below shows the path of a tennis ball in flight. At point, W, 1.5 m above the ground, the tennis ball is travelling upwards at 13 m s^{-1} and angle 60° to the horizontal. The ball rises to a maximum height at point, X, before passing through point Y, which is at the same height as W. It lands at point Z.

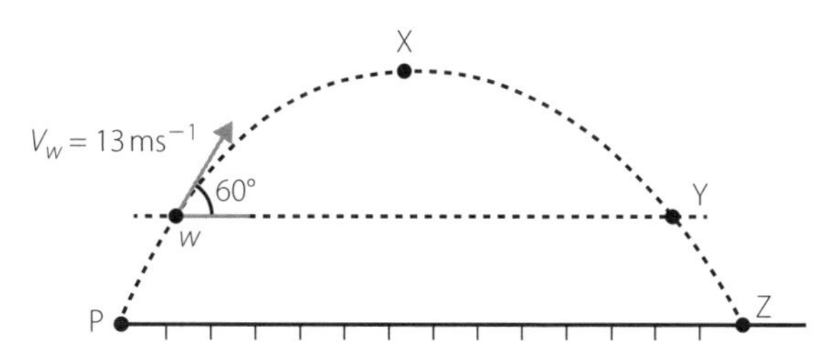

Calculate the:

a horizontal component of the velocity at W

b vertical component of the velocity at Y

c time taken to reach Y

d maximum height attained by the ball

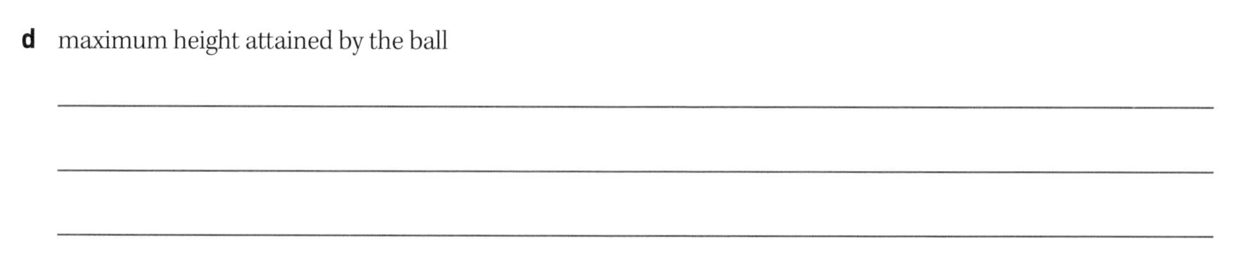

e distance WY

f vertical component of the velocity at Z

g speed at Z

h angle to the horizontal of the velocity at Z

i time taken to reach Z

j horizontal distance between X and Z.

3 In a movie, a character is required to jump across a 2.0 m gap between two buildings of the same height. The character launches at a speed of 4.9 m s^{-1} and angle 23°.

a Show that the character does not succeed in landing on top of the building opposite.

b If the character just crashes through a window below the top of the opposite building, what is the minimum distance from the top of the building to the top of the window?

4 A ball is kicked a distance of 30.0 m over level ground at an initial speed of 8.0 m s^{-1}. At what angle was the ball kicked?

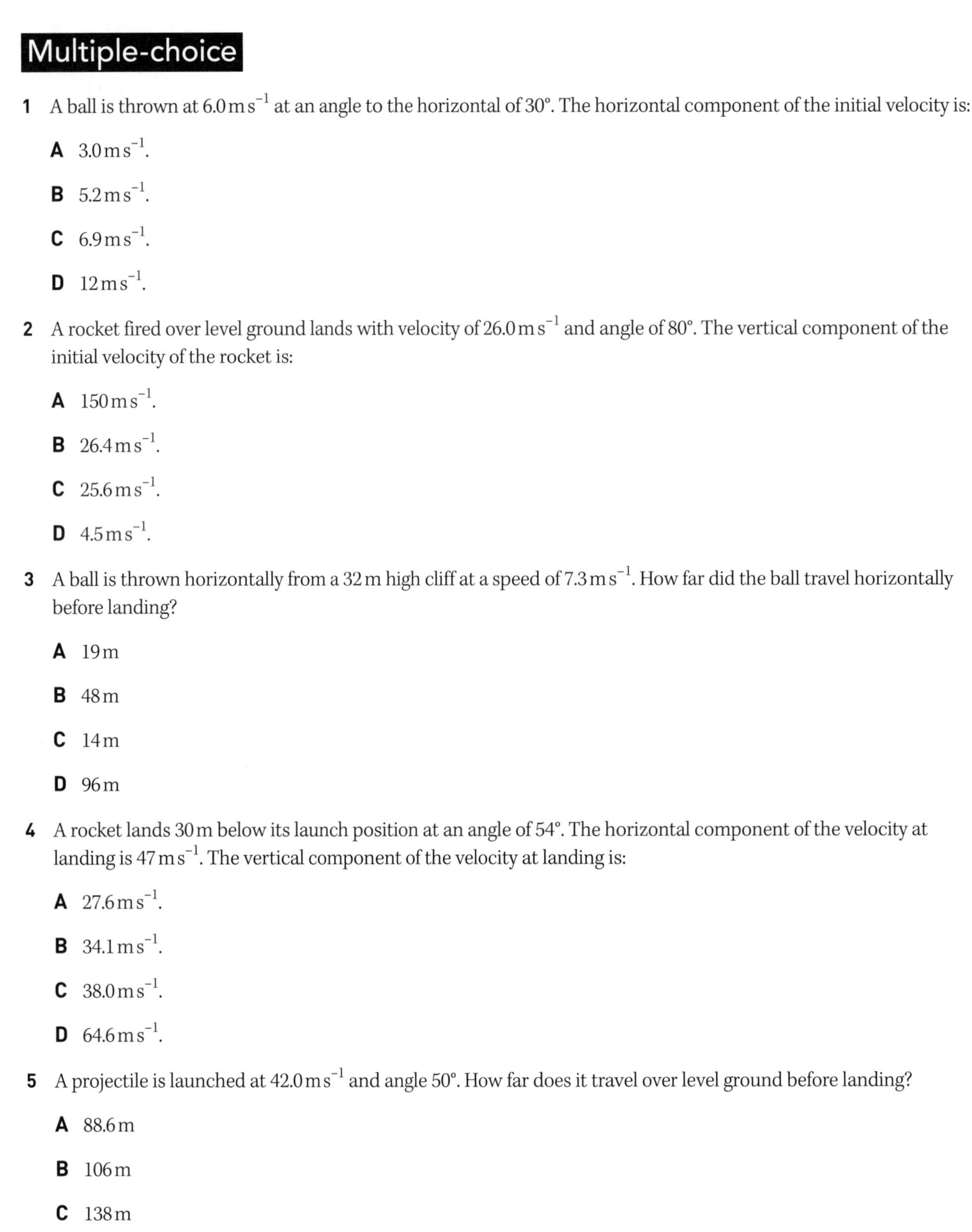

Multiple-choice

1 A ball is thrown at $6.0\,m\,s^{-1}$ at an angle to the horizontal of 30°. The horizontal component of the initial velocity is:

A $3.0\,m\,s^{-1}$.

B $5.2\,m\,s^{-1}$.

C $6.9\,m\,s^{-1}$.

D $12\,m\,s^{-1}$.

2 A rocket fired over level ground lands with velocity of $26.0\,m\,s^{-1}$ and angle of 80°. The vertical component of the initial velocity of the rocket is:

A $150\,m\,s^{-1}$.

B $26.4\,m\,s^{-1}$.

C $25.6\,m\,s^{-1}$.

D $4.5\,m\,s^{-1}$.

3 A ball is thrown horizontally from a 32 m high cliff at a speed of $7.3\,m\,s^{-1}$. How far did the ball travel horizontally before landing?

A 19 m

B 48 m

C 14 m

D 96 m

4 A rocket lands 30 m below its launch position at an angle of 54°. The horizontal component of the velocity at landing is $47\,m\,s^{-1}$. The vertical component of the velocity at landing is:

A $27.6\,m\,s^{-1}$.

B $34.1\,m\,s^{-1}$.

C $38.0\,m\,s^{-1}$.

D $64.6\,m\,s^{-1}$.

5 A projectile is launched at $42.0\,m\,s^{-1}$ and angle 50°. How far does it travel over level ground before landing?

A 88.6 m

B 106 m

C 138 m

D 177 m

6 A ball is thrown at a 30° angle from the top of one building to the top of an identical building, which is 18.0 m away. What was the minimum launch speed?

A 352 m s^{-1}

B 204 m s^{-1}

C 19.0 m s^{-1}

D 14.3 m s^{-1}

7 At launch, a rocket has a speed of 845 m s^{-1}. The rocket returns to ground 15.2 km away. What was the launch angle?

A 24.7°

B 12.4°

C 6.0°

D 3.0°

8 A ball is thrown at 42° and 15.0 m s^{-1} from the outfield from a height of 2.5 m. At its highest point, the ball is 4.0 m above the ground. It is caught 0.5 m from the ground. How long was the ball in the air?

A 1.68 s

B 2.22 s

C 2.31 s

D 11.3 s

9 A stone, launched at 30°, takes 15 s to travel 28 m before landing 38 m below the launch point. The vertical component of the initial speed was:

A 3.7 m s^{-1}.

B 3.2 m s^{-1}.

C 1.1 m s^{-1}.

D 0.93 m s^{-1}.

10 A projectile travels 72 m horizontally in 8.0 s. At this time its speed was 24 m s^{-1}. The angle at which the projectile was travelling is:

A 68°.

B 66°.

C 22°.

D 21°.

11 The horizontal and vertical components of a projectile at launch are, respectively, 15.3 m s^{-1} and 24 m s^{-1}. The speed and angle relative to the horizontal of the projectile at the top of the flight are, respectively:

A 15.3 m s^{-1}; 0°.

B 15.3 m s^{-1}; 32.5°.

C 28.5 m s^{-1}; 0°.

D 28.5 m s^{-1}; 32.5°.

12 A cricket ball struck from the pitch at a speed of $25\,m\,s^{-1}$ lands 45 m away in 3.0 s. At what angle did it leave the pitch?

A 89°

B 53°

C 37°

D 31°

13 A projectile is aimed so as to strike a target 105 m above the launch position. It is launched at a speed of $72.0\,m\,s^{-1}$ and angle 60°. The time taken to reach the target is:

A 58.0 s.

B 14.2 s.

C 10.7 s.

D 2.00 s.

14 A child attempts to jump across a 1.5 m stream at the beach. The maximum speed the child can run is $3.4\,m\,s^{-1}$. The maximum distance the child can jump is:

A 2.0 m.

B 1.5 m.

C 1.2 m.

D unable to be calculated due to insufficient data.

15 Two balls, P and Q, are launched from the same position simultaneously. Ball P is dropped while ball Q is projected horizontally at $6.0\,m\,s^{-1}$. A little later, P has a speed of $5.0\,m\,s^{-1}$. How far from P is Q?

A $3.1\,m\,s^{-1}$

B $4.0\,m\,s^{-1}$

C $1.3\,m\,s^{-1}$

D $2.6\,m\,s^{-1}$

16 A stone is thrown downwards at an angle of 30° to the horizontal with a speed of $20.0\,m\,s^{-1}$. After 2.5 s, the stone has fallen downwards 55.7 m and the speed of the stone is:

A $35.9\,m\,s^{-1}$.

B $38.6\,m\,s^{-1}$.

C $43.0\,m\,s^{-1}$.

D $45.3\,m\,s^{-1}$.

17 During take-off, a light aeroplane is travelling at $9.0\,m\,s^{-1}$ at an angle of 20° when a packet is thrown out horizontally forwards at $3.0\,m\,s^{-1}$. The packet takes 2.5 s to reach the ground. How far horizontally did the packet travel?

A 7.5 m

B 22.5 m

C 28.6 m

D 30.0 m

18 When it is 75.0 m above ground, a helicopter is rising vertically at 14.0 m s^{-1}. At this moment, a parcel is dropped from the helicopter. What is the maximum height reached by the parcel?

A 271 m

B 196 m

C 85 m

D 75 m

19 A light aeroplane is flying horizontally with a speed of 20 m s^{-1} when it releases a package of medical supplies. The point of release of the package is 700 m above the ground. How far does the aeroplane travel before the package reaches the ground?

A 351 m

B 39 m

C 175 m

D 98 m

20 An amusement park trolley is travelling in a vertical circle of radius 25 m at a speed of 8.0 m s^{-1} when it leaves the track. At this point it has climbed to two-thirds of the height of the circle. What is the horizontal velocity component of the speed at the point where the trolley leaves the track?

A 8 m s^{-1}

B $8 \times \sin 60°$ m s^{-1}

C $8 \times \cos 30°$ m s^{-1}

D $8 \times \cos 60°$ m s^{-1}

Short answer

21 A baseball player strikes a baseball 1.5 m above the ground. It leaves the baseball bat at an angle of 50° and is caught 40 m away by a fielder at a height of 1.5 m from the ground.

a Calculate the initial speed of the baseball.

____

____

____

____

____

b Determine the time it took for the baseball to reach the fielder.

____

____

____

____

If the baseball were caught in the outfield and 2.5 m above the ground, find the:

c time taken for the ball to travel to the fielder

d speed of the ball when it is caught.

22 A light aircraft is travelling horizontally at 11.0 m s^{-1} at an altitude of 200 m. A small gizmo, part of the light aircraft, explodes off the back and is projected initially horizontally to the rear at a speed relative to the aircraft of 3.0 m s^{-1}. For a person observing the gizmo from the ground, find the:

a initial horizontal component of the gizmo's velocity

b time taken for the gizmo to reach the ground

c horizontal distance travelled by the gizmo before it hits the ground

d velocity of the gizmo on reaching the ground.

23 A ball is propelled upwards from the top of a 40 m high cliff at a speed of 8.0 m s^{-1} and angle 30°. Simultaneously, a rocket is propelled upwards at an angle of 60° and speed 32 m s^{-1} from a location that is 120 m horizontally from the base of the cliff. Both move in a position coordinate system with origin at the base of the cliff. Ignore air resistance. Assume $g = 9.8$ m s^{-2}.

a Find the maximum time that the ball could spend in the air.

b Find the maximum possible range of the rocket.

c Find the position (x, y), relative to the base of the cliff, at which the ball and rocket collide.

24 A golfer strikes a ball across a water hazard towards a hole, which is horizontally directly opposite. The golf ball is launched at an angle of 70° and speed 36 m s^{-1}. It lands next to the hole. Calculate the:

a distance from tee to hole

b total time the golf ball is in the air

9780170412643

c maximum height reached by the golf ball

d time for which the golf ball is higher than 50 m above the position of the tee.

3 Inclined plane

LEARNING

Summary

- Motion along an inclined plane is principally analysed using Newton's laws.
  - Newton's first law: an object will travel at constant velocity unless affected by a non-zero net external force, ΣF.
  - Newton's second law: the rate of change of velocity, the acceleration, a, can be quantified in terms of the net force and the mass, m: $a = \frac{\Sigma F}{m}$; therefore, $\Sigma F = ma$.
- For real surfaces, three forces that always apply are:
  - $\vec{w}$ = weight (vertically down)
  - $\vec{N}$ = normal force (perpendicular to surface)
  - $\vec{f}$ = friction (parallel to surface).
- In the case of frictionless surfaces, $\vec{f} = 0$.
- On an inclined plane the angle of inclination, θ: $0° < \theta < 90°$.
- For an object on an inclined plane, the net force, $\Sigma\vec{F}$, is the vector sum of normal, weight and friction forces: $\Sigma\vec{F} = \vec{N} + \vec{w} + \vec{f}$; therefore, $\vec{N} + \vec{w} + \vec{f} = m\vec{a}$.
- For motion on an inclined plane, resolve the vector equation into its rectangular components:
  - parallel to the plane
  - perpendicular to the plane.

  This analytic tool reduces equations of motion to simple, algebraic sums, one perpendicular to the surface and one parallel to the surface.
- Resolutes of weight:
  - component of weight perpendicular to the surface:

$$w_{\perp} = w\cos\theta$$
$$\Rightarrow w_{\perp} = mg\cos\theta$$

  - component of weight parallel to the surface:

$$w_{\parallel} = w\sin\theta$$
$$\Rightarrow w_{\parallel} = mg\sin\theta.$$

- The component perpendicular to the surface is the magnitude of the normal force, because there is no acceleration in the direction perpendicular to the plane. By Newton's second law:

$$\Sigma F_{\perp} = 0$$
$$\Rightarrow N - mg\cos\theta = 0$$
$$\Rightarrow N = mg\cos\theta.$$

- Friction is the effect of one surface on another. It is caused by interactions between surfaces.
- For objects sliding, without rolling, there are two types of friction:
  - static friction, f_s: prevents motion occurring; rises to a maximum, beyond which motion begins; and depends on the normal force applied by the surface on the object
  - kinetic friction, f_k : approximate constant force that opposes motion, once motion has begun; and usually less than static friction.
- On an inclined plane without friction:
  - there is no acceleration in the direction perpendicular to the plane:

$$\Sigma F_{\perp} = 0$$
$$\Rightarrow N - mg\cos\theta = 0$$
$$\Rightarrow N = mg\cos\theta$$

  - there will be acceleration in the direction parallel to the plane because, for a frictionless plane, there is no static friction and no kinetic friction

$$\Sigma F_{\parallel} = ma$$
$$\Rightarrow w_{\parallel} - f = ma$$
$$\Rightarrow mg\sin\theta - f = ma$$
$$\text{but } f = 0 (f_s = 0 \text{ and } f_k = 0)$$
$$\Rightarrow a = g\sin\theta.$$

- On an inclined plane with friction:
  - when a force is applied to an object, static friction must be overcome. Static friction rises to a maximum as the force applied to the object increases. Once the object is moving, the kinetic friction remains constant, despite increases in the applied force. Kinetic friction is typically less than the maximum static friction
  - there is no acceleration in the direction perpendicular to the plane:

$$\Sigma F_{\perp} = 0$$
$$\Rightarrow N - mg\cos\theta = 0$$
$$\Rightarrow N = mg\cos\theta$$

  - friction is affected by the normal force. The less the normal force applied by the plane to the mass, the less the friction force on the mass. The normal force applied to a particular mass on an inclined plane decreases as the angle of inclination goes from 0° to 90°, that is, from $N = mg$ when $\theta = 0°$ ($\cos\theta = 1$) to $N = 0$ when $\theta = 90°$ ($\cos\theta = 0$)
  - the normal force may be increased by increasing the component of the force perpendicular to the surface, for example, by pushing on the object
- The maximum static friction, f_s , can be deduced by finding the angle, θ, at which the object begins to accelerate. Up to this point, the acceleration parallel to the surface is zero:

$$mg\sin\theta - f_s = 0$$
$$\Rightarrow f_s = mg\sin\theta$$

  - Once accelerating parallel to the slope, the (constant) kinetic friction, f_k is involved:

$$\Sigma F_{\parallel} = ma = mg\sin\theta - f_k$$
$$\Rightarrow mg\sin\theta - f_k = ma.$$

3.1 Resolving forces

WORKED EXAMPLES

A 10.0 kg mass is on a steep slope of 60° to the horizontal. Find the:

1 component of the weight force parallel to the surface

2 component of the weight force perpendicular to the surface

3 normal force by the surface on the mass.

ANSWERS

1 $w_{||} = mg\cos 60°$

$\Rightarrow w_{||} = 10\,\text{kg} \times 9.8\,\text{m}\,\text{s}^{-2} \times \sqrt{\frac{3}{2}}$

$\Rightarrow w_{||} = 849\,\text{N}$

2 $w = mg\cos 60°$

$\Rightarrow w = 10\,\text{kg} \times 9.8\,\text{m}\,\text{s}^{-2} \times \frac{1}{2}$

$\Rightarrow w = 49\,\text{N}$

3 $N - mg\cos 60° = 0$ (Newton's second law)

$\Rightarrow N = 5.0\,\text{N}$

QUESTIONS

1 Find the net force on a 75 kg mass, which accelerates down an 11° slope at $1.8\,\text{m}\,\text{s}^{-2}$.

____

____

____

____

2 **a** For a 5.0 kg mass on a slope of 30° to the horizontal, find:

i the force on the mass perpendicular to the horizontal

____

____

____

ii the component of the gravitational force parallel to the surface

iii the component of the gravitational force perpendicular to the surface.

b Explain why the normal force and the component of the gravitational force perpendicular to the surface are equal and opposite, but not an action-reaction pair of forces.

3 Complete the table below.

ANGLE (°)	MASS (kg)	WEIGHT (N)	$w_{\parallel}$ (N)	$w_{\perp}$ (N)	NORMAL FORCE (N)
20	10				
30	15				
	20				180
60		250			
			472	626	

4 If the only forces acting on a 23 kg mass on a surface are its weight and the normal force, at what rate will its speed change as a result of being lifted to 8.6° above the horizontal?

3.2 Friction and motion on an inclined plane

WORKED EXAMPLES

1 Find the acceleration of a 10 kg mass, which slides on a frictionless surface inclined at 30°.

2 A 2.0 kg mass begins to move down a surface when it is raised to an angle of 10°. When the angle of the slope is increased to θ it accelerates at 4.0 m s^{-2} and the surface applies a 2.0 N frictional force on the mass.

a Calculate the static friction, f_s.

b Calculate the angle, θ, of the slope.

c Find the normal force when the object is accelerating at 4.0 m s^{-2}.

ANSWERS

1 $a = g\sin\theta$

$\Rightarrow a = 9.8\,\text{m}\,\text{s}^{-2} \times \sin 30°$

$\Rightarrow a = 4.9\,\text{m}\,\text{s}^{-2}$

2 **a** Parallel to surface (about to move – static friction):

$\Sigma F_{||} = f_s - mg\sin\theta = 0$

$\Rightarrow f_s = mg\sin\theta$

$\Rightarrow f_s = 2.0\,\text{kg} \times 9.8\,\text{m}\,\text{s}^{-2} \times \sin 10°$

$\Rightarrow f_s = 3.4\,\text{N}$

b Parallel to surface (moving $\Rightarrow$ kinetic friction):

$\Sigma F_{||} = mg\sin\theta - f_k = ma$

$$\Rightarrow \sin\theta = \frac{ma + f_k}{mg}$$

$$\Rightarrow \sin\theta = \frac{2.0\,\text{kg} \times 4.0\,\text{m}\,\text{s}^{-2} + 2.0\,\text{N}}{2.0\,\text{kg} \times 9.8\,\text{m}\,\text{s}^{-2}}$$

$\Rightarrow \sin\theta = 0.5102$

$\Rightarrow \theta = 31°$

c Perpendicular to surface (no acceleration in perpendicular direction):

$\Sigma F_{\perp} = N - mg\cos\theta = 0$

$\Rightarrow N = mg\cos\theta$

$\Rightarrow N = 2.0\,\text{kg} \times 9.8\,\text{m}\,\text{s}^{-2} \times \cos 31°$

$\Rightarrow N = 17\,\text{N}$

QUESTIONS

1 A 325 g mass is stationary on a plane while it is lifted gently up. At 15° the mass begins to move with an acceleration of $0.84\,\text{m}\,\text{s}^{-2}$. Find:

a the static friction

b the kinetic friction

c the normal force

9780170412643

2 A 0.56 kg mass is subject to a kinetic friction force of 4.0 N as it accelerates from rest down an inclined plane at $2.3\,m\,s^{-2}$. Find:

a the angle of the plane

b the speed after 3.2 s.

3 On a 30° snow slope a 90 kg skier begins to slide. If the kinetic friction is 85% of the static friction, with what acceleration does the skier slide down the slope?

4 A person applies 50 N of force perpendicular to a 40° inclined plane on a 60 kg mass. This causes the friction force to increase to 40 per cent of the magnitude of the normal force. Find:

a the component of the weight force parallel to the surface

b the normal force

c the acceleration of the mass.

5 In order to find how friction changes, a 4.5 kg box was placed on an inclined plane. The plane was slowly raised. The graph below shows how the friction changed during the experiment.

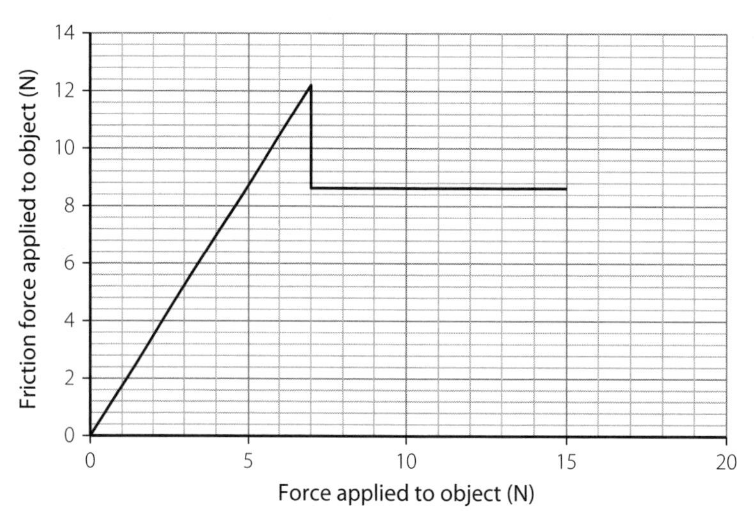

a Explain the shape of the graph.

b At what angle of inclination does the box begin to slide?

c Find the acceleration of the box on a:

i 15° slope

ii 50° slope.

9780170412643

3.3 Solution procedure: inclined plane

WORKED EXAMPLE

A rope is attached to a block of mass 100 kg to keep it from moving on a plane which is inclined at 22° to the horizontal. The rope makes an angle of 10° above the surface of the plane. The friction between the plane and the surface is 850 N. Find the:

1 component of the weight force perpendicular to the surface

2 component of the weight force parallel to the surface

3 component of the force applied by the rope parallel to the surface

4 tension in the rope

5 normal force.

ANSWER

1 $w_{\perp} = mg\cos\theta$

$\Rightarrow w_{\perp} = 100\,\text{kg} \times 9.8\,\text{m s}^{-2} \times \cos 22°$

$\Rightarrow w_{\perp} = 909\,\text{N}$

2 $w_{||} = mg\sin\theta$

$\Rightarrow w_{||} = 100\,\text{kg} \times 9.8\,\text{m s}^{-2} \times \sin 22°$

$\Rightarrow w_{||} = 367\,\text{N}$

3 $\Sigma F = w_{||} - T_{||} = 0$

$\Rightarrow T_{||} = w_{||}$

$\Rightarrow T_{||} = 367\,\text{N}$

4 $T_{||} = T\cos 10° = 367\,\text{N}$

$\Rightarrow T = \dfrac{367\,\text{N}}{\cos 10°}$

$\Rightarrow T = 373\,\text{N}$

5 $N + T_{\perp} - w_{\perp} = 0$

$\Rightarrow N = w_{\perp} - T_{\perp}$

$\Rightarrow N = 909\,\text{N} - T\sin 10°$

$\Rightarrow N = 909\,\text{N} - 373\sin 10°$

$\Rightarrow N = 844\,\text{N}$

QUESTIONS

1 A 4.5 g mass is on a steep slope of 75° to the horizontal. It is just about to move.

a Find the:

i component of the weight force parallel to the surface

____

____

____

ii component of the weight force perpendicular to the surface

iii normal force by the surface on the mass.

b If the kinetic friction is 90% of the static friction, at what rate does the mass slide down the slope?

2 A mass of 10.0 kg is held stationary on a 30° frictionless slope by a rope, which is parallel to the slope.

a Sketch the vector sum of the forces applied to the mass.

b Find the resolutes of each force:

i parallel to the slope

ii perpendicular to the slope.

9780170412643

c Explain why the tension in the rope must be 49 N.

3 A 1.2 t tractor travelling at 7.2 km h^{-1} encounters a 56 m section of mud-covered hillside and slides to the bottom in 5.6 s. The friction between the slope and the tractor is 1.0×10^3 N. Find the:

a acceleration of the tractor on the muddy slope

b speed of the tractor at the end of the slide

c angle of inclination of the mud-covered hillside.

4 A 40 kg mass on a frictionless slope accelerates at a rate of 4.0 m s^{-2}.

a Find the angle of the slope.

A force is now applied to the mass to slow its acceleration to $2.5\,m\,s^{-2}$. This force is directed at 10° above the plane.

b Calculate the magnitude of the extra force applied.

5 A 450 g mass is placed on a surface, which is then raised. Just past the angle at which it begins to slide, the mass accelerates at $0.61\,m\,s^{-2}$. The kinetic friction is 0.5 N.

a Find the:

i angle of the plane

ii maximum static friction.

b The angle of inclination of the plane is now changed to 15° and the mass released from rest. Find the time it takes to reach a speed of $3.1\,m\,s^{-1}$.

9780170412643

Multiple-choice

1 For an object on an inclined plane, the normal force and the component of the weight force perpendicular to the surface:

A can be added because they are an action-reaction pair of forces

B can be added because they affect the same object.

C cannot be added because they affect different objects.

D cannot be added because they are different types of forces.

2 For a 20 kg box on a flat surface, which is inclined at 20°, the magnitude of the normal force is nearest to:

A 197 kg m s^{-2}.

B 195 kg m s^{-1}.

C 190 N.

D 68 N.

3 A 20 g mass on a 60° frictionless surface has an acceleration of:

A 0.17 m s^{-2}.

B 1.7 m s^{-2}.

C 0.10 m s^{-2}.

D 1.0 m s^{-2}.

4 A 100 kg box accelerates down a plank of maximum length 10 m, at 4.0 m s^{-2}. It reaches the lower end in 1.1 s. The plank exerts a constant friction force of 300 N. The net force on the box is:

A 680 N.

B 410 N.

C 400 N.

D impossible to tell because no angle of inclination is given.

5 The normal force applied to a 250 N weight on a 60° slope is:

A 12 250 N.

B 1250 N.

C 1225 N.

D 125 N.

6 A vertically upwards force is applied by a string attached to a 20 kg mass, which is sliding at constant speed down a 35° inclined plane. The tension in the string is:

A $\leq 196\,\text{N}$

B $\leq 196 \times \cos 35°$

C $\leq 196 \times \sin 35°$

D $\leq 196 \times \tan 35°$

7 A student performs an experiment into friction by gently lifting up an inclined plane until a 600 g mass begins to move. This occurs at 18°. The student measured:

A kinetic friction = 5.9 N.

B kinetic friction = 1.8 N.

C static friction = 5.9 N.

D static friction = 1.8 N.

8 A 60 kg mass on a plane inclined at 30° is pushed by a force of 25 N, which is directed perpendicular and into the surface. The normal force applied to the mass is:

A 613 N.

B 534 N.

C 509 N.

D 484 N.

9 A 5.0 kg mass accelerates at $2.5\,\text{m}\,\text{s}^{-2}$ down an inclined plane. The friction is 3.0 N. The angle of inclination of the plane is:

A 79°.

B 72°.

C 18°.

D 11°.

10 A string pulls parallel to a 30° slope so that a 10 kg mass accelerates at $1.3\,\text{m}\,\text{s}^{-2}$. If the total frictional forces amount to 40 N, the tension in the string is:

A 81 N.

B 68 N.

C 36 N.

D 22 N.

11 A string attached to a 75 g mass is used to slow an object that is sliding on a frictionless plane. When the plane is inclined at 60°, and the string is vertical, the acceleration is $0.11\,\text{m}\,\text{s}^{-1}$ down the slope. The net force on the object is:

A 8.42×10^{-3} N.

B 8.25×10^{-3} N.

C 6.37×10^{-2} N.

D 6.68×10^{-2} N.

12 A rigid rod is used to push a 40 kg mass up a 20° slope. If the friction amounts to 75% of the weight, the minimum force applied by the rod is:

A 428 N.

B 294 N.

C 160 N.

D 134 N.

13 A 90 kg box is attached to a rope and pulled up a 48° slope with an acceleration of 2.0 $m s^{-2}$. Friction is 35 N. The tension in the rope is closest to:

A 870 N.

B 440 N.

C 215 N.

D 145 N.

14 A 2.0 kg mass is sliding at 2.0 $m s^{-1}$ on a frictionless plane inclined at 30°. The component of the total force parallel to the plane is:

A 9.8 N, down the slope.

B 9.8 N, up the slope.

C 19.6 N, up the slope.

D 0 N.

15 A 60 g object accelerates down a slope at 1.2 $m s^{-2}$. The slope applies a friction force of 0.10 N. The angle of the slope is:

A 83°.

B 73°.

C 17°.

D 7°.

16 A 4.0 kg block is pushed up a 30° frictionless slope by a force of 24.0 N. The acceleration of the block is:

A 1.1 $m s^{-2}$.

B 1.9 $m s^{-2}$.

C 4.6 $m s^{-2}$.

D 6.0 $m s^{-2}$.

17 A force, *F*, is applied to a block of mass 3.0 kg on a 30° rough surface. The block is accelerating at 1.3 $m s^{-2}$. The kinetic friction force on the block is 6.0 N. The magnitude of the force, *F*, is:

A 24.9 N.

B 14.7 N.

C 12.6 N.

D 3.9 N.

18 A 5.0 kg mass is about to slide down a plane, which is angled at θ to the horizontal. The static friction force amounts to 20% of the normal force. The angle, θ is:

A 63°.

B 16.5°.

C 11.3°.

D 1.38°.

19 A 12 N force is applied to an 8.0 kg mass on a 30° plane up the slope. The frictional force applied by the plane to the mass is 10% of the normal force applied to the mass. The acceleration of the mass down the plane is:

A 2.3 m s^{-2}.

B 2.6 m s^{-2}.

C 4.9 m s^{-2}.

D 7.2 m s^{-2}.

20 A 2.0 kg mass is attached to a 3.0 kg mass. They are placed on a slope which is angled at 30° to the horizontal. A force, F, is applied to the 3.0 kg mass such that the system accelerates at a rate of 2.5 m s^{-2}. The total frictional force on the system amounts to 14.0 N. What is the magnitude of the force, F?

A 20.0 N

B 24.5 N

C 38.5 N

D 51.0 N

Short answer

21 A 6.0 g block slides down a frictionless plane, which is inclined at 17° to the horizontal. A sketch space has been provided below.

a Calculate the:

i normal force

9780170412643

ii weight force parallel to the surface

iii acceleration.

b A string is now attached to the mass and held vertical such that the object travels at $2.0\,m\,s^{-1}$. Calculate the tension in the string. Use Newton's Laws to justify your answer.

22 A 15 N force is applied at 15° above a plane in order to push a 10 kg mass up a plane, which is inclined at 35° to the horizontal. The kinetic friction is 5.0 N. A sketch space has been provided below.

a Find the normal force applied by the plane.

b Calculate the net force applied to the mass.

c Find the acceleration of the mass.

23 A 2.5 kg mass is raised slowly on a plane. It begins to slide when the angle is 22°. For any angle, the kinetic friction is 40% of the normal force. A sketch space has been provided below.

a At the point where the mass begins to slide, find the:

i normal force

ii static friction

iii kinetic friction.

b How long will it take for the mass to travel 8.0 m down the slope when the angle is increased to 30°?

24 Two, 5.0 kg masses are connected over a frictionless pulley. One mass hangs vertically, while the other mass is on a plane inclined at 24° to the horizontal. Assume the friction between the plane and the mass is a constant 15 N. A sketch space has been provided below.

a Find the net force down the slope.

b Calculate the acceleration of the system.

c Find the the angle, θ, for which the system is balanced.

4 Circular motion

Summary

- Uniform circular motion of a point particle is described in terms of the radius of the circle, r, and the time taken for the particle to complete one revolution, T, called the period.
  - Period, T (s) is time taken for one revolution.
  - Frequency, f (hertz; Hz; s^{-1}) is the number of times an object completes one revolution of the circle in a given time interval.
- For questions requiring the conversion of period to frequency or frequency to period follow the steps below.
  - Read the question carefully
  - Identify the quantity provided
  - Identify the appropriate equation:
  - $f = \frac{1}{T}$ or $T = \frac{1}{f}$
  - Substitute values, including units
  - Solve
  - Check to ensure the question has been answered in the correct units
- For uniform circular motion, the speed at any point is the same; hence the average speed for one revolution is the same as the instantaneous speed at any point on the circle. Therefore, from the definition of average speed:

$$v = \frac{\text{distance traveled}}{\text{time interval}}$$

$$\Rightarrow v = \frac{2\pi r}{T} = 2\pi r f$$

- Uniform circular motion is accelerated motion:
  - speed is constant
  - velocity changes
  - the acceleration is directed towards the centre of the circle (centripetal or centre seeking).
- Linking acceleration with measured or given quantities:
  - acceleration given speed and period or frequency:

$$a = \frac{2\pi v}{T}$$

$$a = 2\pi v f$$

9780170412643

- acceleration given radius and period or frequency:

$$a = \frac{4\pi^2 r}{T^2} \text{ OR } a = 4\pi^2 r f^2$$

- acceleration given speed and radius:

$$a = \frac{v^2}{r}$$

- From Newton's second law, an acceleration of a mass is caused by the sum of all real forces acting on the object, $\Sigma F(\text{on object})$.
  - Acceleration is caused by real forces: normal, friction, tension, electric, magnetic and gravitational.
  - The sum of forces that cause uniform circular motion is also referred to as centripetal force.
- The net (centripetal) force:
  - acts in the same direction as the acceleration
  - is the vector resultant of one or more real forces
  - is not a real force, although it may be the resultant of only one real force.
- The centripetal force is not a new kind of force – it is the *sum* of all real forces applied to an object moving at constant speed around a circle. It is *not* a separate kind of real force.

- Linking net force with mass, speed and radius:

$$\Sigma F = m\frac{v^2}{r}$$

- Linking net force with mass, radius and period or frequency:

$$\Sigma F = m\frac{4\pi^2 r}{T^2}$$

$$\Sigma F = m \times 4\pi^2 r f^2$$

- On a banked circular track or roadway the three real forces that act on the object are:
  - normal force: the normal force is always at right angles to the surface
  - friction: parallel to the surface ($f = 0$, negligible in many practical cases).
  - weight: perpendicular to ground.
- For a mass travelling in a horizontal circle on a banked track, the net force on the mass is horizontal. It is the vector sum of the normal force and the weight force.

$$\Sigma F = mg\tan\theta = m\frac{v^2}{r}$$

$$\Rightarrow a = g\tan\theta = \frac{v^2}{r}$$

- The assumption that f is negligible means that it is the limiting case between:
  - sliding up the slope (speed too fast; f down the slope)
  - sliding down the slope (speed too slow; f up the slope)
- Motion in a vertical circle can be quite complicated to analyse; however, forces at the upper and lower positions are relatively straightforward. Consideration must be given to whether the mass is on the inside or outside of the track on which it is moving:
  - At the top:

$$\Sigma F = \{\text{sum of downward forces}\} - \{\text{sum of upward forces}\} = \frac{mv^2}{r}$$

  - At the bottom:

$$\Sigma F = \{\text{sum of upwards forces}\} - \{\text{sum of downwards forces}\} = \frac{mv^2}{r}$$

- Object on the end of a string:
  - at the top: $T + w = \frac{mv^2}{r}$
  - at the bottom: $T - w = \frac{mv^2}{r}$.

- For a car travelling over a rise, ΣF is downwards, in the direction of the weight force:

$$\Sigma F = w - N = m\frac{v^2}{r}$$

$$\Rightarrow N = mg - m\frac{v^2}{r}$$

When $N = 0$, the car leaves the road at speed:

$$v_{max} = \sqrt{gr}$$

- For a car travelling through a dip:

$$\Sigma F = N - w = m\frac{v^2}{r}$$

$$\Rightarrow N = m\frac{v^2}{r} + mg$$

9780170412643

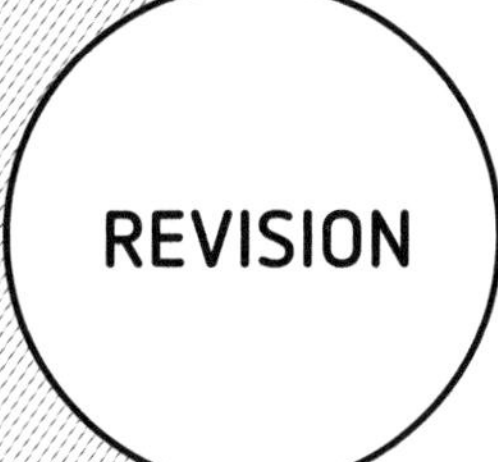

4.1 Uniform circular motion

WORKED EXAMPLES

1 A small ballbearing takes 2.0 s to complete 10 revolutions at constant speed on a horizontal circle of radius 3.0 cm. Calculate the:

a frequency of rotation

b speed of the ballbearing.

2 A rubber stopper is whirled on a horizontal circle of radius 60 cm at a speed of $1.8\,\text{m}\,\text{s}^{-1}$. Calculate the period of revolution.

ANSWERS

1 a $f = \dfrac{10 \text{ revolutions}}{2.0\text{ s}}$

$f = 5.0\text{ s}^{-1}$

b $v = 2\pi rf$

$\Rightarrow v = 2\pi \times (3.0 \times 10^{-2}\text{ m}) \times 5.0\text{ s}^{-1}$

$\Rightarrow v = 0.94\text{ m s}^{-1}$

2 $v = \dfrac{2\pi r}{T}$

$\Rightarrow T = \dfrac{2\pi r}{v}$

$\Rightarrow T = \dfrac{2\pi \times 0.60\text{ m}}{1.8\text{ m s}^{-1}}$

$\Rightarrow T = 2.1\text{ s}$

QUESTIONS

1 Compare period and frequency.

____

____

2 Complete the table below.

r (m)	T (s)	f (s^{-1})	v (m s^{-1})
1.5	2.0		
	20		4.0
0.60			15.0

3 A 1.8 m radius revolving door completes 5.0 revolutions in 44.0 s. Calculate the:

a frequency of rotation

b speed of a point on the edge of the door.

4 A funpark swing carousel consists of a chair on a wire chain, which revolves at a rate of 12 turns in 1.0 minute on a horizontal circle of radius 6.20 m. Calculate its:

a period

b speed.

5 A plane flying horizontally at 240 km h^{-1} describes a circle in 52.0 s. Find the diameter of the circle.

4.2 Centripetal acceleration and force

WORKED EXAMPLES

1 An object travelling at 0.60 m s^{-1} takes 0.45 s to complete one revolution of a circle of radius 2.0 m. Find the acceleration.

2 Find the acceleration of a body revolving on a circle of radius 20 m with a period of 25 s.

3 Find the acceleration of a car travelling on a bend of radius 90 m at 35 m s^{-1}.

4 The acceleration of a satellite is 2.45 m s^{-2} at a radius of 1.27×10^7 m from the centre of Earth. Calculate the:

a frequency of its revolution

b speed of the satellite.

ANSWERS

1 $a = \frac{2\pi r}{T}$

$\Rightarrow a = \frac{2\pi \times 6.0\,\text{m s}^{-1}}{0.45\,\text{s}}$

$\Rightarrow a = 8.4\,\text{m s}^{-2}$

2 $a = \frac{4\pi r^2}{T^2}$

$\Rightarrow a = \frac{4\pi \times (20\,\text{m})^2}{(25\,\text{s})^2}$

$\Rightarrow a = 8.0\,\text{m s}^{-2}$

3 $a = \frac{v^2}{r}$

$\Rightarrow a = \frac{(35\,\text{m s}^{-1})^2}{90\,\text{m}}$

$\Rightarrow a = 13.6\,\text{m s}^{-2}$

4 **a** $a = 4\pi^2 r f^2$

$\Rightarrow f = \sqrt{\frac{a}{4\pi^2 r}}$

$\Rightarrow f = \frac{1}{2\pi}\sqrt{\frac{a}{r}}$

$\Rightarrow f = \frac{1}{2\pi}\sqrt{\frac{2.45\,\text{m s}^{-2}}{1.27 \times 10^{-7}\,\text{m}}}$

$\Rightarrow f = 7.0 \times 10^{-7}\,\text{Hz}$

b $a = 2\pi v f$

$\Rightarrow v = \frac{a}{2\pi f}$

$\Rightarrow v = \frac{2.45\,\text{m s}^{-2}}{2\pi \times 7.0 \times 10^{-7}\,\text{Hz}}$

$\Rightarrow v = 5.6 \times 10^{3}\,\text{m s}^{-1}$

QUESTIONS

1 Complete the table below.

r (m)	v (m s^{-1})	a (m s^{-2})
1.5	4.0	
3.5		9.6
	15.0	14

2 Complete the table below.

r (m)	T (s)	f (s^{-1})	a ($m\,s^{-2}$)
63	40	(–)	
3.5	(–)	0.20	
	19	(–)	8.3
1.2×10^{-3}	(–)		0.47

3 The centre of the Moon is 3.84×10^5 km from Earth's centre. The Moon takes 27.3 days to orbit Earth. Find the Moon's:

a acceleration

____

____

____

b orbital speed.

____

____

____

4 Mars' orbital radius and speed are, respectively, 2.28×10^8 km and $2.41 \times 10^3\ m\,s^{-1}$. Calculate the:

a acceleration of Mars towards the Sun

____

____

____

b the period of Mars' orbit around the Sun.

____

____

____

5 The acceleration due to gravity some distance from Earth is $2.6\ m\,s^{-2}$. At this distance, a satellite makes 6.32 passes over a city every day. Calculate the orbital speed of the satellite.

____

____

____

9780170412643

4.3 Net force causes circular motion

WORKED EXAMPLES

1 A 60 kg marathon athlete goes around a corner of radius 15 m at a speed of $5.5\,\text{m}\,\text{s}^{-1}$. Calculate the net force applied to the athlete.

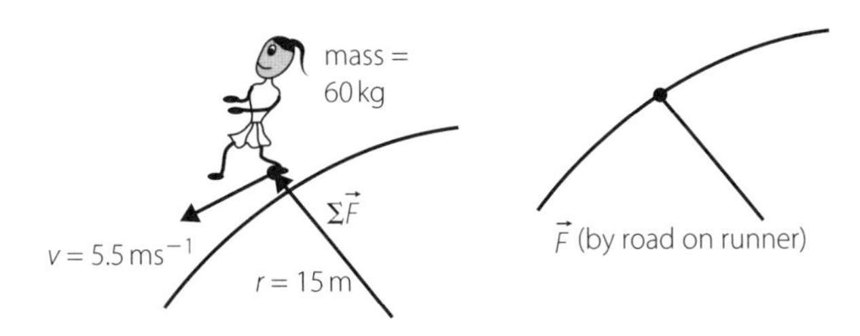

2 A string has a maximum tension of 25 N. What is the maximum period with which a 50 g mass can be made to circulate on a horizontal table at a radius of 55 cm?

ANSWERS

1 $\Sigma F = m\dfrac{v^2}{r}$

$$\Rightarrow \Sigma F = 60\,\text{kg} \times \frac{(5.5\,\text{ms}^{-1})^2}{20\,\text{m}}$$

$$\Rightarrow \Sigma F = 91\,\text{N}$$

2

$m = 50\,\text{g}$
$\Sigma\vec{F}$
$r = 55\,\text{cm}$
O

$$\Sigma F = m\frac{4\pi^2 r}{T^2}$$

$$\Rightarrow T = 2\pi\sqrt{\frac{mr}{\Sigma F}}$$

$$\Rightarrow T = 2\pi\sqrt{\frac{5.0\times10^{-2}\,\text{kg}\times0.55\,\text{m}}{25\,\text{N}}}$$

$$\Rightarrow T = 0.21\,\text{s}$$

QUESTIONS

1 A funpark rotor has a radius of 7.0 m and frequency of 0.20 Hz. A 60 kg person is in contact with the wall as it rotates.

a Identify the force that causes the person to rotate.

b Calculate the net force applied to the person.

2 A net force of 67 N is needed to enable a 150 g mass to revolve on the end of a string in a horizontal circle of radius 35 cm. Calculate the time it takes for the mass to complete one revolution.

3 A 450 kg motorbike travels around a horizontal curve of radius 65 m at 95 km h^{-1}. Find the net force applied by the motorbike on the road.

4 The electrostatic force on an electron travelling in a circular orbit around a proton in a (very simplified) Bohr hydrogen atom is 8.2×10^{-8} N. The radius of the orbit is 5.3 nm. Calculate the frequency with which the electron orbits the proton. Mass of an electron = 9.1×10^{-31} kg

5 A 200 g object is suspended from the ceiling by a string of length 0.80 m. It moves around a horizontal circular path describing 20 revolutions in 23.2 s. The string makes an angle of 30° to the vertical. Find the magnitude of the net force acting on the mass.

4.4 Uniform circular motion on non-horizontal surfaces

WORKED EXAMPLES

1 A 1.5 kg bucket is whirled in a vertical circle on the end of a string of length 1.2 m at a constant speed of 2.1 m s^{-1}. Calculate the tension in the string at the bottom of the circle.

2 A 300 kg motorcycle travels horizontally around a road, which is banked at 5°, at a speed of 20 m s^{-1}. Friction is negligible. Calculate the net force acting on the cyclist.

3 A 1.5 t truck travels at speed, 25 m s^{-1}, over a hump in a road. If the hump has a radius of 75 m calculate the:

a normal force applied by the road to the car

b maximum safe speed for driving over the hump.

ANSWERS

1

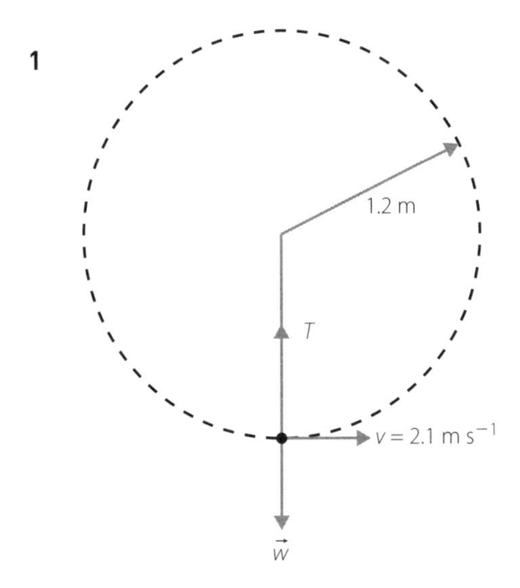

$$\Sigma F = T - mg = m\frac{v^2}{r}$$

$$\Rightarrow T = m\frac{v^2}{r} + mg$$

$$\Rightarrow T = 1.5\,\text{kg} \times \frac{\left(2.1\,\text{m s}^{-1}\right)^2}{1.2} + 1.5\,\text{kg} \times 9.8\,\text{m s}^{-2}$$

$$\Rightarrow T = 20.2\,\text{N}$$

2

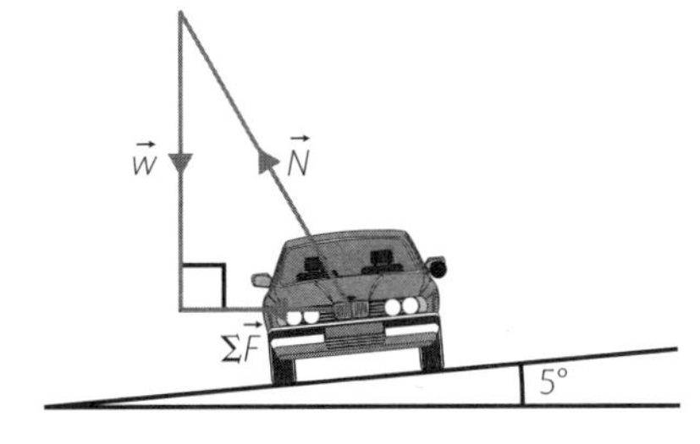

$$\Sigma\vec{F} = \vec{N} + \vec{w}$$

$$\Rightarrow \Sigma F = mg\tan\theta$$

$$\Rightarrow \Sigma F = 300\,\text{kg} \times 9.8\,\text{m s}^{-2} \times \tan 5°$$

$$\Rightarrow \Sigma F = 257\,\text{N}$$

3

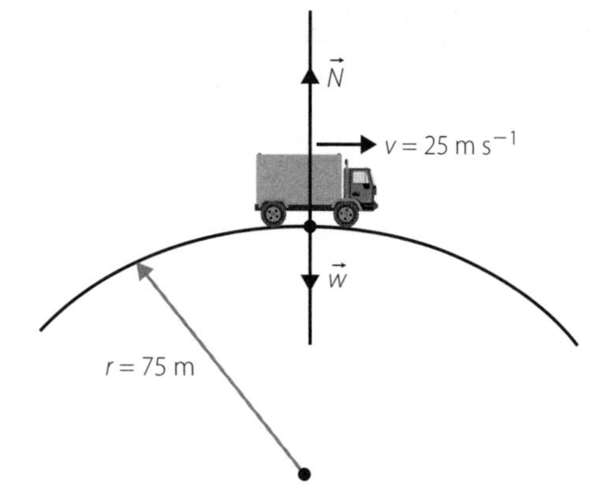

a $\Sigma F = mg - N = m\frac{v^2}{r}$

$\Rightarrow N = mg - m\frac{v^2}{r}$

$\Rightarrow N = 1.5 \times 10^3 \text{ kg} \times 9.8 \text{ m s}^{-2} - 1.5 \times 10^3 \text{ kg} \times \frac{(25 \text{ m s}^{-1})^2}{75 \text{ m}}$

$\Rightarrow N = 2.2 \times 10^3 \text{ N}$

b $v = \sqrt{gr}$

$\Rightarrow v = \sqrt{9.8 \text{ m s}^{-2} \times 75 \text{ m}}$

$\Rightarrow v = 27.1 \text{ m s}^{-1}$

QUESTIONS

1 A 0.35 kg object is suspended from the ceiling by a string, which makes an angle of 68° to the vertical. The mass moves in a horizontal circular path. Calculate the magnitude of the net force acting on the mass.

2 A crane moves a 450 kg load at constant speed on a 48 m cable in a horizontal circular path of radius 14 m. Calculate the speed of the load.

3 A magnetic force of 4.3×10^{-15} N applied continuously at right angles to an ion of mass 4.1×10^{-26} kg travelling horizontally at a speed of $8.0 \times 10^4 \text{ m s}^{-1}$. Find the radius of curvature of the path described by the electron.

4 A 45 kg child is in a small amusement park rotor of radius 5.6 m. The rotor takes 20 turns in 16 s. Find the force applied by the wall of the rotor on the child.

5 A 60 kg rider travels around in horizontal path of radius 50 m on a velodrome banked at 22° at a speed of $10\,\text{m s}^{-1}$

a Calculate the net force on the rider.

b Determine whether or not friction is involved in this scenario.

6 An 80 g mass is swung in a vertical circle on the end of a string of length 65 cm at a constant speed of $7.3\,\text{m s}^{-1}$. Calculate the tension in the string at the:

a top of the circle

b bottom of the circle.

7 A 105 kg cyclist travels over a speed hump of radius 2.1 m at a speed of $25\,\text{km h}^{-1}$. Decide whether the cyclist leaves the road or not. Use calculations to support your answer.

8 While driving through a dip in the road of radius 25 m, a 95 kg driver experiences a force that is three times the usual gravitational force. How fast was the driver going?

EVALUATION

Multiple-choice

1 In SI units, period is measured in:

A s.

B s^{-1}.

C Hz.

D rpm.

2 A dust particle on the edge of a vinyl record completes 33 revolutions per minute. The frequency, in SI units, is:

A 0.55 Hz.

B 0.55 s.

C 1.8 s.

D 33 rpm.

3 Calculate the time it takes for a cyclist travelling at $12\,m\,s^{-1}$ to go once around a circular horizontal track of radius 60 m.

A 130 s

B 10π s

C 1.3 s

D 31 s

4 A ball is attached to a string of length 1.32 m and whirled around in a circle. The angle between the string and a vertical line through the centre of the circle is 15°. It takes 20 s for the ball to rotate 35 times. The speed of the ball is:

A $4.7\,m\,s^{-1}$.

B $3.8\,m\,s^{-1}$.

C $1.2\,m\,s^{-1}$.

D $0.37\,m\,s^{-1}$.

5 A 45 kg person in an amusement park ride travels in a horizontal circle of radius 7.0 m. What is the acceleration of the person when they are travelling at a rate of 0.5 revolutions per second?

A $138\,m\,s^{-2}$

B $13.8\,m\,s^{-2}$

C $69\,m\,s^{-2}$

D $6.9\,m\,s^{-2}$

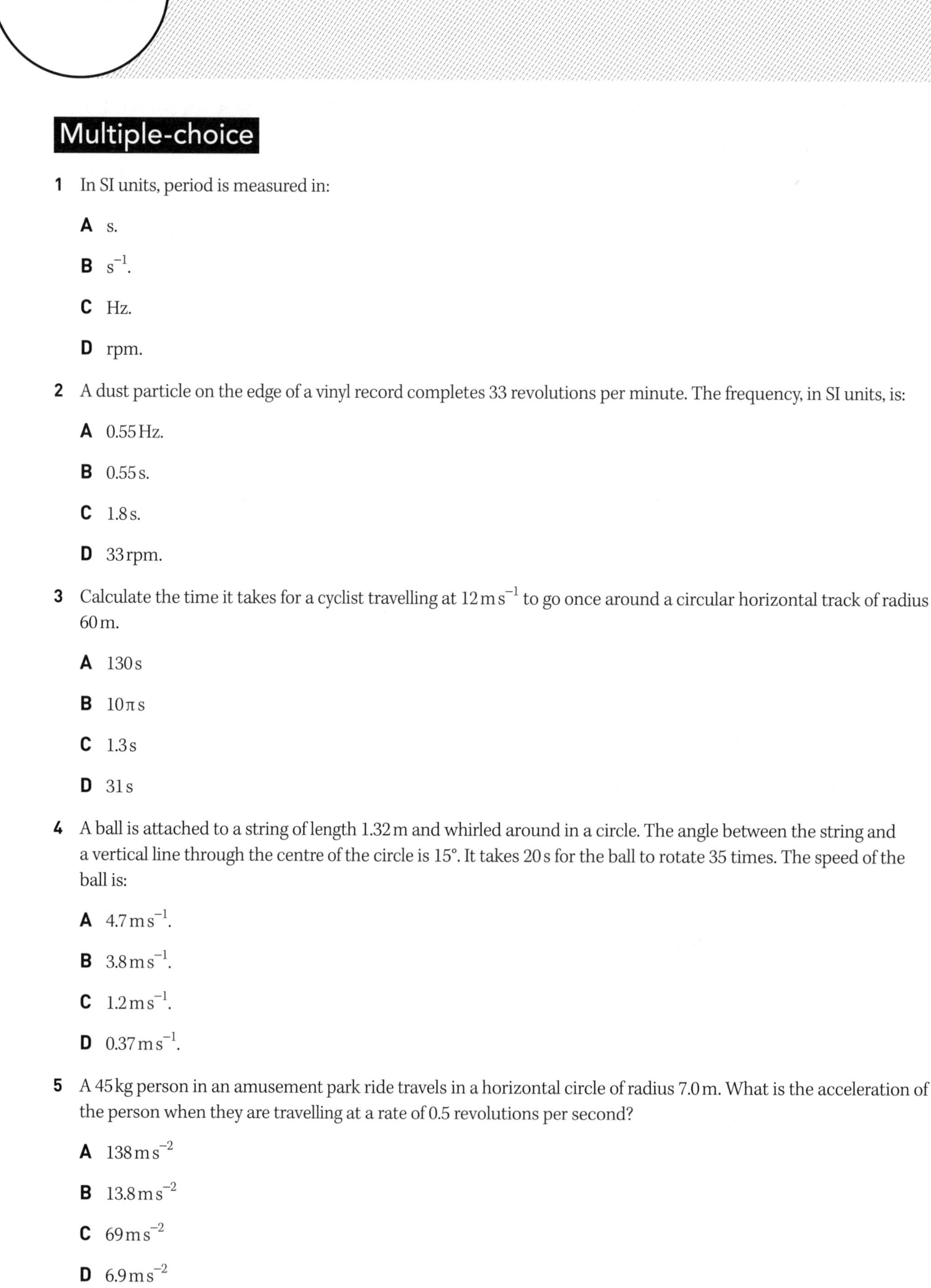

9780170412643

6 A magnetic field causes an electron to travel in a circle of radius 40 cm when it has a speed of $2.0 \times 10^7\ \text{m s}^{-1}$. The acceleration of the electron is:

A $1.0 \times 10^{-15}\ \text{m s}^{-2}$.

B $1.6 \times 10^{8}\ \text{m s}^{-2}$.

C $1.0 \times 10^{15}\ \text{m s}^{-2}$.

D $6.3 \times 10^{-15}\ \text{m s}^{-2}$.

7 A cyclist travels around a velodrome with a banked surface. The force or forces that keep the cyclist travelling in a horizontal circle are:

A the weight force and the normal force.

B the weight force.

C the normal force.

D the centripetal force and the net force.

8 A 1.0 t truck travels around a horizontal bend of radius 40 m at a speed of $20\ \text{m s}^{-1}$. The magnitude of the net force by the truck on the road is:

A 0.50 N.

B 5.0×10^2 N.

C 1.0×10^3 N.

D 1.0×10^4 N.

9 An object undergoes uniform circular motion in a circle of fixed radius at speed, v, because of a net force, F. If the speed of the object is doubled, the net force required is:

A $0.25F$.

B $2F$.

C $4F$.

D $8F$.

10 A 10 g mass is attached to a 50 cm long string and swung in a vertical circle. When the mass is at the top of the circle, the tension in the string is zero. Therefore, the speed of the mass at the top is:

A zero.

B $2.2\ \text{m s}^{-1}$.

C $7.0\ \text{m s}^{-1}$.

D $9.8\ \text{m s}^{-1}$.

11 The ability to steer a car while travelling through a dip in the road is:

A easier, because the upwards centripetal force on the car reduces the downwards weight force.

B easier, because the upwards normal force on the car reduces the downwards weight force.

C harder, because the downwards reaction force by the car to the centripetal force increases the total downwards force.

D harder, because the downwards reaction force by the car to the upwards normal force increases the total downwards force.

12 A 100 kg cyclist travels at a speed of $10\,m\,s^{-1}$ horizontally around a road, which is banked at 10° to the horizontal. Friction is negligible. Calculate the net force acting on the cyclist.

A 170 N

B 173 N

C 965 N

D 5.5 kN

13 A train travels horizontally at a speed of $360\,km\,h^{-1}$ on a curve, which is banked. The radius of the track is 360 m. The maximum angle of banking for the train to travel around the curve safely is:

A 1.6°

B 0.6°

C 1.4°

D 7.1°

14 A 970 kg vehicle travels around a bend, which is banked at an angle of 8° to the horizontal, at a speed of $20\,m\,s^{-1}$. The force applied by the truck perpendicular to the surface of the road is:

A 133 N.

B 1.3×10^3 N.

C 9.4×10^3 N.

D 6.6×10^4 N.

15 What radius of curvature is required for a 2.0 kg mass to travel at $3.0\,m\,s^{-1}$ in a horizontal circular path around a slope banked at 20° to the horizontal?

A 1.6 m

B 2.5 m

C 3.3 m

D 5.0 m

16 Calculate the period of revolution of an ion of mass 3.0×10^{-27} kg, which is subjected to a net force of 9.0×10^{-18} N acting at right angles to its circular path of radius 10 cm.

A $3.6 \times 10^{-5}\,s^{-1}$

B 2.8×10^4 s

C $5.8 \times 10^{-7}\,s^{-1}$

D 1.7×10^6 s

17 Calculate the net force on a 10.0 kg mass at the equator of Earth, which has a radius of 6.38×10^6 m.

A 0.34 N

B 9.8 N

C 65 N

D 98 N

18 A truck has a mass of 2.5×10^3 kg. It travels over a speed hump of radius 3.0 m at a speed of $15\,m\,s^{-1}$. What happens to the truck?

A The truck stays on the road because it is travelling at a speed less than the speed needed for the normal force to become zero.

B The truck stays on the road because the weight is large enough to force the truck down onto the road.

C The truck leaves the road because the normal force on the truck is large enough to force the truck up from the road.

D The truck leaves the road because it is travelling at a speed greater than the speed necessary for the normal force to become zero.

19 A 50 g mass is being swung in a vertical circle on the end of a string. It takes 20 s for it to make 50 revolutions. When it is halfway going up to the top of the circle its speed is $4.0\,m\,s^{-1}$ and the string breaks. What happens to the mass?

A The mass continues on at $4.0\,m\,s^{-1}$ upwards and slows to a stop directly above the point of release.

B The mass starts travelling outwards at $4.0\,m\,s^{-1}$, directly away from the centre, because the centrifugal force is greater than the centripetal force.

C The mass starts travelling inwards at $4.0\,m\,s^{-1}$, directly towards the centre, because the centripetal force is greater than the centrifugal force

D The mass begins to drop directly down, gaining speed because gravity is greater than the centripetal force.

20 When a runner changes course on a circular path, the force that causes the motion to change is applied:

A on the ground towards the centre of the circle.

B on the ground away from the centre of the circle.

C by the ground towards the centre of the circle.

D by the ground away from the centre of the circle.

Short answer

21 The figure below shows a small, spherical glass paperweight of mass 200 g suspended from the ceiling by a 80 cm long string. It is swinging in a circle of radius 16 cm. Ignore all resistance forces. Assume $g = 9.8\,\text{m}\,\text{s}^{-2}$.

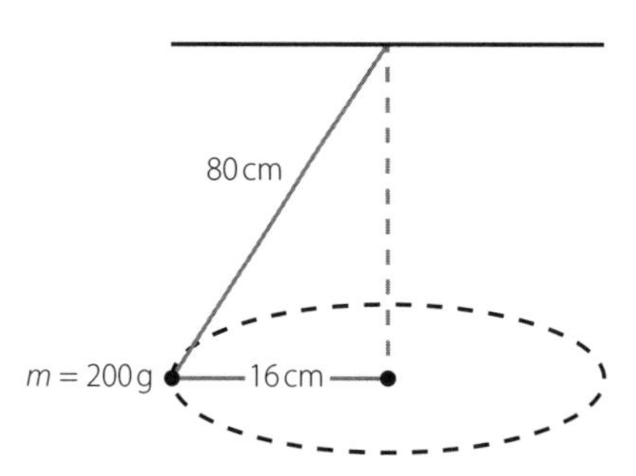

a Draw a free-body diagram to represent all the forces acting on the glass paperweight.

b Calculate the net force on the ball.

c Find the tension in the string.

d Deduce the frequency of the ball's motion.

9780170412643

22 The drawing below shows a 60 kg ice-skater performing a circular, gliding movement around a 30° arc of a circle of radius 4.5 m. Her skate blade is shown. It makes an angle, θ, to the vertical, while she travels with a speed of $3.0\,m\,s^{-1}$. Ignore all resistance forces. Assume $g = 9.8\,m\,s^{-2}$.

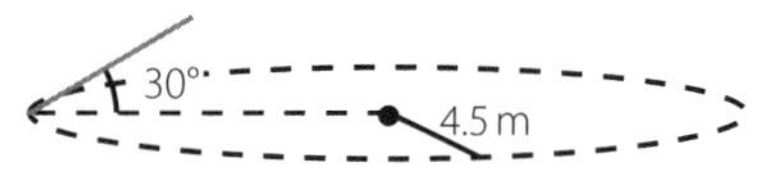

a On the diagram, show the direction of the net force on the skate blade.

b How long does it take the skater to complete the movement?

____

____

____

c Calculate the magnitude of the net force on the skater.

____

____

____

d Find the angle, θ.

____

____

____

23 A small steel ball of mass 200 g has been allowed to slide freely around a vertical, circular ball race of diameter 1.0 m. Ignore all resistance forces. Assume $g = 9.8\,m\,s^{-2}$.

a What is the speed of the steel ball at the bottom of the ball race, such that it experiences a force that is five times its weight?

____

____

____

b What minimum speed must be reached so that the ball maintains contact with the ball race?

____

____

____

c Show that, for a ball that is free to go around the ball race, the net force on the ball is not directed towards the centre, except at the top and at the bottom.

____

____

____

24 A 1000 kg vehicle travels horizontally around a slope, which is at 18° to the horizontal.

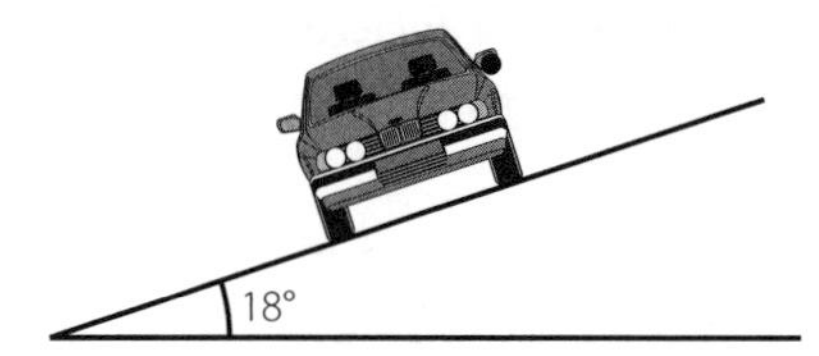

a Calculate the magnitude of the normal force by the surface on the vehicle.

b Find the net force, $\Sigma\vec{F}$, on the vehicle.

c Find the radius, r, when the vehicle is travelling at $8.0\,\text{m}\,\text{s}^{-1}$.

25 A 250 kg motorcycle now travels around the same slope as Question **24** at radius $\frac{r}{2}$

a What is the acceleration of the motorcycle?

b How fast is the motorcycle travelling?

5 Gravitational force and field

Summary

- Gravity is an action-at-a-distance force. Action-at-a-distance forces can be modelled as being mediated by a field.
- The gravitational field at a particular point is defined as the force per unit mass experienced by an object at that point: $g = \frac{F}{m} = \frac{GM}{r^2}$.
- The gravitational field has units of $\mathrm{N\,kg^{-1}}$ or $\mathrm{m\,s^{-2}}$.
- In field theory, we model the energy as being stored in the field.
- When an object moves in a gravitational field, work is done on or by the field.
- When work is done by the field, the potential energy of the system decreases. When work is done on the field, the potential energy of the system increases.
- Potential energy is the energy stored in a system or field. Potential energy gives a system the ability to do work. Gravitational potential energy is such an example.
- Work is the energy transferred due to the action of a force: $W = Fs$.
- Early models of the universe had the heavenly bodies fixed on spheres in space, including crystalline spheres and later epicycles, to explain the observed motion of the planets.
- Galileo observed that falling bodies had constant horizontal motion and accelerated vertical motion.
- Newton's law of universal gravitation, $F = G\frac{Mm}{r^2}$, relates the mass of objects with the distance between them using the Newtonian constant of gravitation, $G = 6.67 \times 10^{-11}\ \mathrm{N\,m^2\,kg^{-2}}$.
- The inverse-square law describes a relationship in which the dependent variable is proportional to the square of the inverse of the independent variable.
- Gravitational force:
  - acts at a distance
  - acts between the centres of an object's mass (their centre of gravity)
  - is calculated as the force per unit mass, $\mathrm{N\,kg^{-1}}$, directed towards the centre of mass of an object
  - like gravitational field, is a vector quantity, having both magnitude and direction.

- The centre of mass is the average position of the mass in an object, or group of objects.
- Force weight is found using Newton's second law: $F_W = mg$. On the surface of Earth g is approximately 9.80 m s^{-2}.

TABLE 5.1 Comparing the four fundamental forces

TYPE	PARTICLE THAT CARRIES THE FORCE	AFFECTS	RELATIVE MAGNITUDE
Electromagnetic	Photons	Charged particles	$\times 10^{36}$
Gravitational	Graviton	Objects with mass	$\times 10^{0}$
Strong nuclear	Gluons	Quarks, gluons	$\times 10^{38}$
Weak nuclear	W^+, W^-, Z^0	Quarks, leptons	$\times 10^{29}$

9780170412643

5.1 Gravitational force and field key terms

QUESTIONS

1 Demonstrate your understanding of gravitational forces and gravitational fields by writing a paragraph relating the key terms listed below.

- Gravity – a force that acts at a distance, attracting all bodies with mass.
- Newton's law of universal gravitation – the relationship between force gravitation, the mass of objects and the distance or radius between the objects.
- Gravitational potential energy – the potential energy associated with the interaction of objects via the gravitational force; potential to do work.
- Newton's second law of motion – relates the acceleration of an object to the force applied and the mass. The acceleration is proportional to the net force applied and inversely proportional to the mass of the object.

2 Draw a flowchart or mind map relating and linking the key terms listed below.

- Inverse-square law – a relationship in which the dependent variable is proportional to the square of the inverse of the independent variable.
- Newton's constant of gravitation – the value of $6.67 \times 10^{-11}\,\text{Nm}^2\,\text{kg}^{-2}$.
- Newton's third law of motion – for every action force, there is an equal and opposite reaction force.
- Newton's law of universal gravitation – the relationship between force gravitation, the mass of objects and the distance or radius between the objects.

5.2 The history of gravity

QUESTIONS

1 State a contribution to the understanding of gravitational acceleration by each of the following scientists.

a Aristotle

b Newton

c Galileo

d Einstein

2 List two formulae that exhibit inverse square relationships.

3 Predict the net difference (increase, decrease, remains the same) for the value of the gravitational force, F, if:

a the mass, M, decreases

____

b the radius, r, decreases

____

c both the radius and the mass double.

____

4 Complete the following statements using most appropriate word from the list below.

Gravitas	Empirical
Kinematics	Galileo
Work	Potential energy

a The energy transferred due to the action of a force is otherwise termed ______________.

b ______________ is the energy stored in a system that gives the system the ability to do work.

c The Aristotelian idea about the 'heaviness' of objects made of earth that allowed them to fall in straight lines towards Earth is termed ______________.

d Experiments on projectiles and the motion of falling objects were carried out in the fifteenth and sixteenth centuries. The most significant of these experiments were undertaken by ______________ who showed that falling objects accelerated uniformly towards Earth.

e The central tenet of the scientific method whereby hypotheses are tested using observation and experimentation is an ______________ method.

f ______________ is the relationship between measurements of distance and time to analyse motion.

5.3 Gravitational potential energy

QUESTIONS

1 Calculate the work is done in raising an 80 kg mass through a vertical height of 18 m.

____

____

____

2 The gravitational potential energy associated with a stationary object is 3.063×10^3 J.

a Determine the height of the object above the ground given it has a mass of 125 kg.

____

____

b Determine the maximum kinetic energy of the object if it were to be dropped to the ground.

____

____

3 **a** How much work is done by the gravitational field when a 75 kg trampolinist falls through a vertical height of 1.8 m?

b Using the conservation of energy, determine what velocity the trampolinist will be travelling at the lowest point.

4 An Atlas rocket of 597 000 kg is launched from ground level. When it is at an altitude of 600 m its vertical velocity is 550 m s^{-1}.

a What is the gravitational potential energy of the rocket at an altitude of 600 m?

b What is the E_k of the rocket when the rocket is at 600 m altitude?

c How much work was done on the rocket to change its gravitational potential and kinetic energy?

5 Work is done on a 1600 kg vehicle to raise it 30 m vertically further up a hill from its initial position.

a Determine the minimum amount of work required to be done to raise the vehicle to its new height up the hill.

b When released from this new, higher position, it rolls to its original position under gravity. What is the maximum velocity of the vehicle at this point, neglecting any other forces acting?

9780170412643

5.4 Gravitational fields

QUESTIONS

1 Calculate the difference in the force weight of an astronaut and their suit, with a combined mass of 150 kg, on the surface of Earth and the surface of Mars. Use:

- $g_{Earth} = 9.80\,m\,s^{-2}$
- $g_{Mars} = 3.69\,m\,s^{-2}$

2 Voyager 1, launched on 5 September 1977, had a launch mass of 825.5 kg. In deep space, beyond our solar system, it's force of weight is negligible. Calculate its weight on Earth prior to launch.

3 The Mars Pathfinder rover, Sojourner, has a mass of 10.5 kg. Determine the force of weight of Sojourner on the Martian surface. Use:

- $g_{Mars} = 3.69\,m\,s^{-2}$

4 The force of gravitational attraction between small bodies is said to be negligible. Determine the gravitational attraction between two people of 60 kg mass at a distance of 2 m to confirm that this force is in fact negligible.

5.5 Gravitational field strength

QUESTIONS

1 A ball is thrown from a window with an initial upward velocity of $2.0\,m\,s^{-1}$. It is found to hit the ground after a period of 2.0 s. Determine how high above the ground the window is.

2 Europa (one of Jupiter's Galilean moons) has a mass of 4.80×10^{22} kg and a radius of 1.56×10^{6} m. Determine the acceleration due to gravity on the surface of Europa.

3 An object is thrown vertically upwards with an initial velocity of $15.0\,m\,s^{-1}$.

a What is the maximum height reached by the object?

b How long will the ball take to fall back to its original position?

4 Compare the gravitational field strengths of the Galilean moons of Jupiter, using the data provided in the table below.

MOON	MASS (kg)	RADIUS (m)
Europa	4.80×10^{22}	1.56×10^{6}
Io	8.93×10^{22}	1.83×10^{6}
Callisto	1.08×10^{23}	2.41×10^{6}
Ganymede	1.48×10^{23}	2.63×10^{6}

9780170412643

5.6 Newton's law of universal gravitation and gravitational force

QUESTIONS

1 Contrast the formulas for gravitational field and gravitational force.

2 Two masses, m and M, have a gravitational force of attraction F when they are a distance r apart. What is the relative magnitude of the gravitational force when the distance is increased to $2r$ and the masses are increased to $3m$ and $2M$?

3 Titan, one of Saturn's moons, is captured in orbit due to the equal and opposite gravitational forces that act between the planet Saturn and Titan. Determine the size of the gravitational force. Use:

- $\text{Mass}_{\text{Saturn}} = 5.67 \times 10^{26}\,\text{kg}$
- $\text{Mass}_{\text{Titan}} = 1.35 \times 10^{23}\,\text{kg}$
- $\text{Mean distance}_{\text{Saturn–Titan}} = 1.22 \times 10^{9}\,\text{m}$.

4 Determine the net gravitational force acting on a spacecraft that is positioned 100 000 km from Earth on a line between Earth and the Moon. Use:

- $\text{Mass}_{\text{Earth}} = 5.97 \times 10^{24}\,\text{kg}$
- $\text{Mass}_{\text{Moon}} = 7.34 \times 10^{22}\,\text{kg}$
- $\text{Mean distance}_{\text{Earth–Moon}} = 3.84 \times 10^{8}\,\text{m}$.

EVALUATION

Multiple-choice

1 The gravitational field model explains why objects may exert forces at a distance. The model also allows us to:

A predict the acceleration of an object within a gravitational field.

B calculate the mass of an object from the observed force that it exerts on another object.

C calculate the mass of distant objects, such as planets, by observing their orbits about a star.

D All of the above

2 The correct units for Newton's universal gravitational constant, *G*, are:

A $\mathrm{N\,m\,kg^{-2}}$

B $\mathrm{N\,kg^{-1}}$

C $\mathrm{m^{3}\,s^{-2}\,kg^{-1}}$

D $\mathrm{N\,m^{2}\,kg^{-2}}$

Short answer

3 Determine the force of weight of the 15 200 kg Apollo lunar lander on the surface of the Moon. Use:

- $\text{Mass}_{\text{Moon}} = 7.35 \times 10^{22}$ kg.

____

____

____

4 Calculate the gravitational field exerted by the Moon on Earth as well as the gravitational field exerted by the Sun on Earth. Explain why the gravitational field due to the Moon is larger than that of the Sun. Use:

- $\text{Distance}_{\text{Earth–Moon}} = 3.84 \times 10^{8}\,\text{m}$
- $\text{Distance}_{\text{Earth–Sun}} = 1.50 \times 10^{11}\,\text{m}$
- $\text{Mass}_{\text{Moon}} = 7.35 \times 10^{22}\,\text{kg}$
- $\text{Mass}_{\text{Sun}} = 1.99 \times 10^{30}\,\text{kg}$
- $\text{Mass}_{\text{Earth}} = 5.97 \times 10^{24}\,\text{kg}$.

5 Contrast the terms gravitational field and gravitational force.

6 Determine the force of gravitational attraction between two objects of 10 kg mass at a distance of 5 m apart. Use calculations to explain what happens to the gravitational force between the two objects if they are moved to twice the distance.

7 Calculate the time of flight of an object of mass x kg that is projected vertically upward from Earth's surface with an initial velocity of $8.0\,m\,s^{-1}$.

8 A 10.0 kg rock is dropped from a height of 5.0 m above the surface of Mercury. Use:
- $Mass_{Mercury} = 3.28 \times 10^{23}$ kg
- $Radius_{Mercury} = 2.57 \times 10^{6}$ km.

a Determine the gravitational field near the surface of Mercury.

b Find the gravitational acceleration of the rock.

c Calculate how much gravitational potential energy was transformed as the rock fell to the surface.

d Calculate the maximum amount of kinetic energy gained by the rock.

9 Calculate the gravitational potential energy of a body of 75 kg raised to a height 2.4 m above Earth's surface.

10 a Calculate the time period, T, for a single oscillation of a pendulum on Earth, given a length of 0.50 m.

b State how the period differs if the experiment were conducted on a planet, such as Neptune, with a greater acceleration due to gravity.

11 Determine the minimum vertical velocity that an object requires to reach a height of 100 m.

6 Orbital motion

LEARNING

Summary

- The geocentric model is a superseded model of the solar system with the Sun, Moon and planets revolving about Earth at its centre.
- The heliocentric model is the current model of the solar system with the Sun (Helios) at its centre and all planets revolving about it. It is closely associated with the work of Nicolaus Copernicus and Galileo Galilee.
- Kepler's first law (the law of ellipses): all planets move in elliptical orbits with the Sun at one focus.
- An ellipse is a regular, curved shape that is a conic section (formed by cutting a cone obliquely). An ellipse describes the path of satellites in orbit around larger bodies.
- Kepler's second law (the law of equal areas): a line that connects a planet to the Sun sweeps out equal areas in equal time periods.
- The mean orbital distance is the average radius of orbit of one massive object about another, such as Earth revolving about the Sun.
- Kepler's third law (the law of periods): the square of the period of a planet's orbit is proportional to the cube of its mean orbital distance. $T^2 \alpha\, r^3$.
- The astronomical unit (AU), is a unit of measure equivalent to Earth's mean orbital radius about the Sun. $1.0\,\text{AU} = 1.50 \times 10^8\,\text{km} = 1.50 \times 10^{11}\,\text{m}$.
- The megaparsec (Mpc), is the distance subtended by an angle of one arc second $\times 1 \times 10^6$. $1.0\,\text{Mpc} = 3.09 \times 10^{19}\,\text{km} = 3.09 \times 10^{22}\,\text{m}$.
- The light year (ly), is a measure of the distance that light would travel in one year. $1.0\,\text{ly} = 9.47 \times 10^{12}\,\text{km} = 9.47 \times 10^{15}\,\text{m}$.
- A satellite is a natural (e.g. the Moon) or synthetic (e.g. GPS or communications satellite) body that orbits a significantly larger mass.
- The force of weight is the force acting on an object due to a gravitational field: $Fw = mg = \frac{GMm}{r^2}$.
- An orbit is a regularly repeated elliptical path of one object about another massive object, such as a planet about a sun.
- Orbital velocity is the precise velocity required for an object to continue to orbit a mass at a given altitude. Orbital velocity: $v = \sqrt{\frac{GM}{r}}$.
- Apparent weightlessness is the experience of having no normal force exerted on you; this occurs during free fall.

9780170412643

- Centre of mass is the average position of the mass in an object, or group of objects. It is the point at which the gravitational force can be modelled as acting when the object is in a gravitational field.
- Geostationary satellites are positioned above one place on Earth. They must travel directly above a point on the equator.
- Geosynchronous satellites complete one orbit of Earth in the same time as Earth completes one revolution.
- Low Earth orbit (LEO) is a satellite orbit within the range of approximately 250 km to 1000 km above Earth's surface.
- Escape velocity is the minimum velocity required for an object to escape the gravitational field of a planet or other large mass.

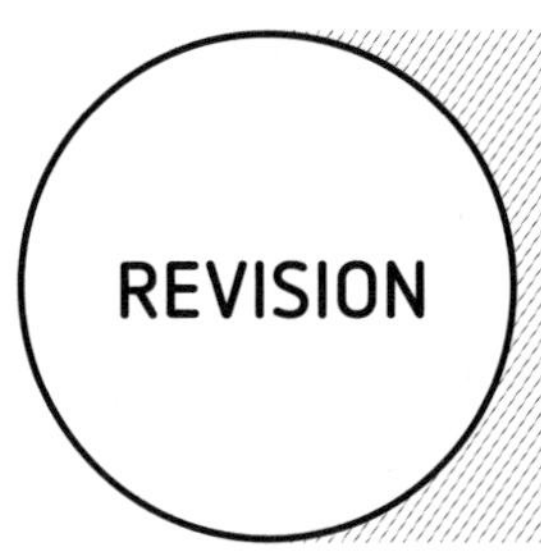

6.1 Orbital motion key terms

QUESTIONS

1 Demonstrate your understanding of orbital motion by writing a paragraph that relates the key terms listed below.

- Kepler's first law (the law of ellipses): all planets move in elliptical orbits with the Sun at one focus.
- Kepler's second law (the law of equal areas): a line that connects a planet to the Sun sweeps out equal areas in equal time periods.
- Kepler's third law (the law of periods): the square of the period of a planet's orbit is proportional to the cube of its mean orbital distance.
- Orbital velocity – the precise velocity required for an object to continue to orbit a mass at a given altitude.
- Apparent weightlessness – the experience of having no normal force exerted on you, as occurs during free fall.

2 Draw a flowchart or mind map that relates and links the key terms listed below.

- Concentric circles – circles that chare a common centre.
- Epicycles – smaller circles whose centre is on the radius of larger circles; used by Ptolemy to describe the motion of planets.
- Geocentric model – a superseded model of the solar system with the Sun, Moon and planets revolving about Earth at its centre.
- Heliocentric model – a current model of the solar system with the Sun at its centre with all planets revolving about it.
- Escape velocity – the minimum velocity required for an object to escape the gravitational field of a planet or other large mass.

9780170412643

6.2 Kepler's laws of planetary motion

QUESTIONS

1 State Kepler's first, second and third laws of planetary motion.

2 Data values for several natural satellites of the Jovian system are shown in the table below. Use the data for the moon Io to determine the ratio of $\frac{R^3}{T^2}$ and hence determine the missing orbital radius and orbital period values of Jupiter's moons Europa, Ganymede and Callisto.

MOON	ORBITAL PERIOD, *t* (days)	ORBITAL RADIUS, *r* (m)
Io	1.78	4.21×10^8
Europa	3.56	
Ganymede		1.06×10^9
Callisto	16.70	

3 The average $\frac{R^3}{T^2}$ value for our solar system is $3.35 \times 10^{18}\,m^3\,s^{-2}$. The mean orbital radius of Jupiter is 7.78×10^{11} m. Use this value, and Kepler's third law, to determine the orbital period of Jupiter.

4 A recently discovered exoplanet, planet C, is found to travel within a nearby galaxy near a known star, star X. It has been suggested that planet C may orbit star X; however, this is yet to be confirmed. Use Kepler's Law and the data provided for planets A and B, that are known to be part of this system, to confirm whether planet C should be classified as part of this system.

	PLANET A	PLANET B	PLANET C (unconfirmed)
Radius in ($\times 10^3$ km)	26.1	9.50	51.2
Mass (kg)	2.56×10^{24}	2.34×10^{24}	3.15×10^{24}
Orbital period (Earth seconds)	8.30×10^6	4.90×10^7	8.84×10^8
Orbital radius (m)	9.04×10^{10}	2.88×10^{11}	20.4×10^{11}
Rotational period (days)	16.66	3.84	1.66

6.3 Newton's law of universal gravitation and Kepler's third law

QUESTIONS

1 Use the formulas for centripetal and gravitational force to derive an equation for the orbital velocity of a satellite.

2 The Pinwheel galaxy in the constellation of Ursa Major is approximately 20.9 million light years from the Milky Way. Convert the distance to the Pinwheel galaxy into:

a kilometres

b megaparsecs

c parsecs.

3 Calculate the orbital period, T, for an artificial satellite orbiting Earth at an altitude of 190 000 km. Use:

- $\text{Mass}_{\text{Earth}} = 5.97 \times 10^{24}\,\text{kg}$
- $\text{Radius}_{\text{Earth}} = 6.37 \times 10^{6}\,\text{m}$.

4 The mean orbital radius of Mercury is 5.79×10^{10} m. Using Kepler's third law our solar system has been found to have an average value of $\frac{T^2}{R^3}$ of approximately $3.41 \times 10^{18}\,\text{s}^2\,\text{m}^{-3}$. Use this value to determine Mercury's orbital period.

6.4 Satellite motion

QUESTIONS

1 Determine the centripetal force acting on Earth by the Sun. Use:

- $\text{Mass}_{\text{Sun}} = 1.99 \times 10^{30}\,\text{kg}$
- $\text{Mass}_{\text{Earth}} = 5.97 \times 10^{24}\,\text{kg}$
- $\text{Distance}_{\text{Sun–Earth}} = 1.50 \times 10^{11}\,\text{km}$.

2 Determine the velocity of a satellite orbiting Earth at an altitude of 4000 km. Use:

- $\text{Mass}_{\text{Earth}} = 5.97 \times 10^{24}\,\text{kg}$
- $\text{Radius}_{\text{Earth}} = 6.37 \times 10^{6}\,\text{m}$.

3 Determine the orbital period, in hours, of a satellite of mass 1250 kg and altitude of 12 000 km above the surface of Earth. Use:

- $\text{Mass}_{\text{Earth}} = 5.97 \times 10^{24}\,\text{kg}$
- $\text{Radius}_{\text{Earth}} = 6.37 \times 10^{6}\,\text{m}$.

____

____

____

4 To what velocity must a satellite be propelled if it is to maintain an orbit with a radius of 500 km around Earth? Use:

- $\text{Mass}_{\text{Earth}} = 5.97 \times 10^{24}\,\text{kg}$
- $\text{Radius}_{\text{Earth}} = 6.37 \times 10^{6}\,\text{m}$.

____

____

____

9780170412643

Multiple-choice

1 The centripetal force and centripetal acceleration that satellites in orbit experience toward Earth is provided by which mechanism?

A Force tension

B Mass

C Force friction

D Force gravity

2 If a satellite's velocity was to be increased, what would be observed about its motion about Earth?

A Its altitude above Earth would decrease.

B Its altitude above Earth would proportionally increase.

C Its altitude above Earth would remain constant.

D Its altitude above Earth would increase, but *not* proportionally.

Short answer

3 What is the orbital period, in Earth years, for an asteroid that is orbiting the Sun with a mean orbital radius that is three times the orbital radius of Earth?

____

____

____

____

4 What is the mass of the central body being orbited by a satellite when the satellite's orbital period is 8.0 hours and its distance from the central body is 4.50×10^8 m?

____

____

____

5 A geostationary satellite orbits Earth once every 24 hours at an altitude of 35 800 km. Determine the value of the Kepler ratio $\frac{R^3}{T^2}$. Use:

- $\text{Mass}_{\text{Earth}} = 5.97 \times 10^{24}$ kg
- $\text{Radius}_{\text{Earth}} = 6.37 \times 10^{6}$ m.

____

____

____

____

6 Contrast the units of length measurement the megaparsec, the light year and the astronomical unit. Which of these units is most useful for measuring the distance across the Milky Way galaxy?

____

____

____

____

7 A natural satellite orbiting the planet Uranus has an orbital radius of 1.29×10^{8} km. Determine the period of revolution for this satellite. Use:

- $G = 6.67 \times 10^{-11}$ N m^2 kg^{-2}
- $\text{Mass}_{\text{Uranus}} = 8.80 \times 10^{25}$ kg.

____

____

____

8 Calculate the gravitational force acting on a satellite of orbital radius 16 500 km and mass 3200 kg. Use:

- $\text{Mass}_{\text{Earth}} = 5.97 \times 10^{24}$ kg
- $\text{Radius}_{\text{Earth}} = 6.37 \times 10^{6}$ m.

____

____

____

9 Determine the escape velocity required for a rocket to escape Mercury's gravitational attraction. Use:

- $G = 6.67 \times 10^{-11}\,\text{N m}^2\,\text{kg}^{-2}$
- $\text{Mass}_{\text{Mercury}} = 3.28 \times 10^{23}\,\text{kg}$
- $\text{Radius}_{\text{Mercury}} = 2.57 \times 10^{6}\,\text{m}$.

---

10 Find the altitude of a satellite in orbit around Earth given that its orbital speed is $8.0\,\text{km s}^{-1}$. Use:

- $\text{Mass}_{\text{Earth}} = 5.97 \times 10^{24}\,\text{kg}$
- $\text{Radius}_{\text{Earth}} = 6.37 \times 10^{6}\,\text{m}$.

---

11 Distinguish between orbital velocity, orbital acceleration and escape velocity.

---

7 Electrostatics

Summary

- First law of electrostatics: like charges repel due to a repulsive force and unlike charges attract due to an attractive force.
- Second law of electrostatics, also known as Coulomb's law: the force exerted between two point charges is proportional to the product of their strengths and inversely proportional to the square of the distances that separates them. $F = \frac{1}{4\pi\varepsilon_0} \times \frac{qQ}{r^2}$ where q and Q are the strengths of the two charges in Coulomb's, r is the distance between the two charges in metres, and ε_0 is the permittivity of free space $\frac{1}{4\pi\varepsilon_0} \approx 9.0 \times 10^9\ \text{Nm}^2\,\text{C}^{-2}$.
- Electrons are free to move through materials. Protons are fixed in the centre of atoms.
- An electric field is exerted by all objects with an overall net charge.
- Electric field lines model net lines of force and point in the direction a positive charge would move in the field. Can only be modelled on stationary charges. More field lines represent a stronger field.
- A uniform field is one pointing in the same direction with the same strength at all points, such as a field between two parallel plate capacitors.
- Electric field strength is quantified by:
  - $E = \frac{F}{q}$ where E is the electric field strength in $\text{N}\,\text{C}^{-1}$ and F is the force in Newtons acting on test charge q in the field
  - $E = \frac{F}{q} = \frac{1}{4\pi\varepsilon_0}\frac{Q}{r^2}$ where E is the electric field strength in $\text{N}\,\text{C}^{-1}$, Q is the charge in C exerting the field and r is the distance from the field source Q in m.
- Electric potential is an intrinsic property of an electric field which models the potential energy per unit charge in that field $V = \frac{U}{q}$, where V is the electron potential in volts (V), U is the potential energy in Joules (J) and q is the size of the charge in the field.

- Potential difference is the difference in electric potential between two points in an electric field $\Delta V = \frac{\Delta U}{q}$ where ΔU is the change in potential energy between two points in the field, also known as the work done on the charge moving in the field.
  - Positive potential difference is when a positive charge moves against the field and hence does work on the field, or when a negative charge moves with the field and hence does work on the field.
  - Negative potential difference is when a positive charge moves with the field and the field does work on the charge, or a negative charge moves against the field and the field does work on the charge.
  - Zero of potential energy is when components in an electric field are infinitely far apart.

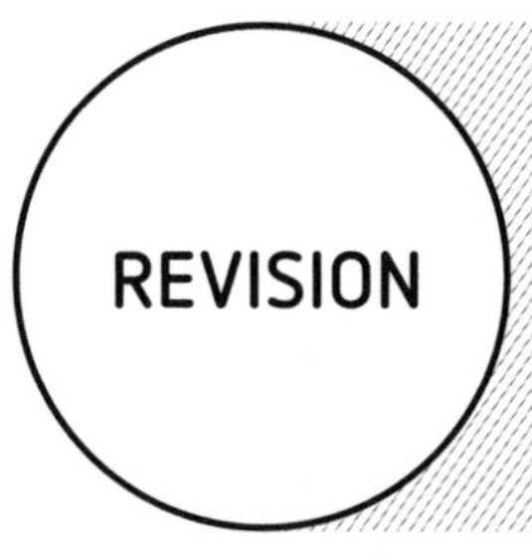

7.1 Coulomb's Law

WORKED EXAMPLE

Calculate the force exerted by a proton on a proton in a helium nucleus at a distance of 6×10^{-11} m.

ANSWER

The charge on a proton is 1.6×10^{-19} C.

$$F = 9.0 \times 10^{9} \times \frac{1.6 \times 10^{-19} \times 1.6 \times 10^{-19}}{\left(6 \times 10^{-11}\right)^{2}}$$

$$F = 6.4 \times 10^{-8}\ \text{N}$$

The positive sign indicates a force of repulsion between the protons.

QUESTIONS

1 State Coulomb's Law.

____

____

2 Compare attractive and repulsive forces.

____

____

3 How is electrostatics different to electricity?

____

____

4 Explain how charge redistributes when two charged objects come into contact with each other.

____

____

5 Name some real life situations where electrostatically charged objects are produced.

____

____

6 Two negative charges A and B are brought into close proximity.

a Draw a vector diagram for this situation.

b What type of force does A experience due to B?

c Does B experience a force? State why (or why not).

d If the size of charge A is now doubled, how does this affect the force that charge B experiences?

e If the distance between A and B was reduced by a factor of $\frac{1}{3}$, how would the force acting on A change?

7 What is the force between two electrons separated by a distance of 10 μm?

8 Two like charges separated by a distance 6.5 μm experience a repulsive force of 62 N. What is the size and magnitude of charge on each of these charges?

9 Calculate the distance between two charges of 5.2 μC and 6.8 μC if the force of repulsion they experience has a magnitude of 33 N.

10 An experiment was conducted where two spheres with 4.1 mC of charge were separated by an increasing distance r. The force of repulsion each charge experienced was measured at each distance r and recorded in the table below.

By plotting a graph of F against $\frac{1}{r^2}$, determine an experimental value for $\frac{1}{4\pi\varepsilon_0}$ from the gradient.

r(m)	F(N)
2	38 000
4	9450
6	4200
8	2410
10	1500
12	1050
14	770
16	600

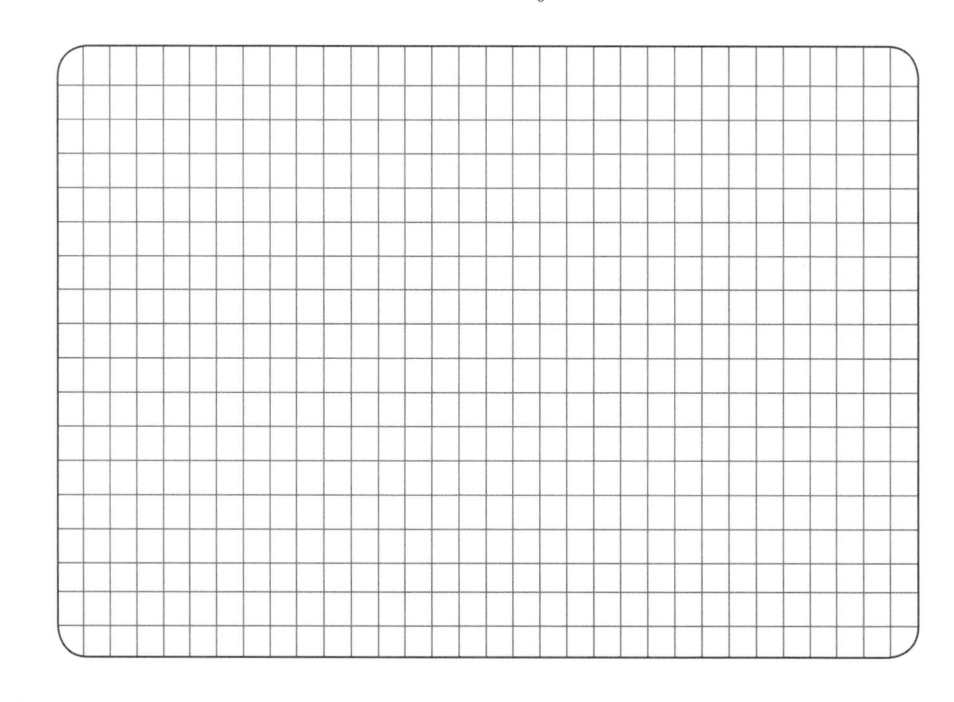

7.2 Electric fields

WORKED EXAMPLE

1 What is the force exerted on an electron when it is experiencing an electric field of 2.3×10^{-3} N C^{-1}?

2 The source emitting this field has a charge of 2×10^{-15} C. How far away was the electron when it was experiencing this field strength from the source?

ANSWERS

1 We know that $E = 3.0$ N C^{-1} and that $q = -1.6 \times 10^{-19}$ C.

$$E = \frac{F}{q}$$

$$F = 2.30 \times 10^{-3} \times -1.6 \times 10^{-19}$$

$$F = -3.68 \times 10^{-22} \text{ N}$$

The force is in the direction opposite to that of the field.

9780170412643

2 We know $E = \frac{1}{4\pi\varepsilon_0} \times \frac{Q}{r^2}$

$$E = 9\times10^{9}\,\text{Cm}^2\,\text{C}^{-2} \times \frac{8\times10^{-6}\,\text{C}}{(0.1\,\text{m})^2}$$

$$E = 7.2\times10^{6}\,\text{NC}^{-1}$$

$$E = \frac{1}{4\pi\varepsilon_0} \times \frac{Q}{r^2}$$

$$E_{1\text{cm}} = 9\times10^{9} \times \frac{6\times10^{-6}}{0.01^2}$$

$$E_{1\text{cm}} = 5.4\times10^{8}\,\text{NC}^{-1}$$

$$E_{10\text{cm}} = 5.4\times10^{6}\,\text{NC}^{-1}$$

$$E_{1\text{m}} = 5.4\times10^{4}\,\text{NC}^{-1}$$

$$r = \sqrt{\frac{1}{4\pi\varepsilon_0}\frac{Q}{E}}$$

so

$$r = \sqrt{9\times10^{9} \times \frac{2\times10^{-15}}{2.3\times10^{-3}}}$$

$$r = \sqrt{0.00783}$$

$$\therefore r = 8.85\times10^{-2}\,\text{m}$$

QUESTIONS

1 Draw the electric field lines coming out of a positive charge.

2 Show how the field lines interact with a negative charge and positive charge when brought into close proximity to each other. Assume the charges are equal in magnitude.

3 Show how electric field lines interact when two like positive charges are brought into close proximity to each other. Assume the charges differ in their magnitude of charge.

4 Explain how you can determine the direction an electron will move with placed in an electric field.

5 A small charge of 4 μC is placed within an electric field. It experiences a force of 3.1×10^{-2} N. Calculate the electric field strength at this point in the field.

6 A point charge $q = 8$ μC generates an electric field. Determine the electric field at a distance of 10 cm from the charge.

7 A charge of −6 μC exerts an electric field E. Compare the field strength at 1 cm, 10 cm and 100 cm away from this charge.

9780170412643

7.3 Electric potential energy

WORKED EXAMPLE

An electron is accelerated through a potential difference of 10 kV before colliding with a target and emitting electromagnetic radiation. What is the velocity of the electron just before it hits the target?

ANSWER

If the electron accelerates from rest, before it hits the target all its potential energy will have been converted into kinetic energy.

$\Delta U = q\Delta V$

$\Delta U = 1.6\times10^{-19}\,\text{C}\times10\,\text{kV}$

$\frac{1}{r^2}$

$\Delta V = q\Delta U$

$\Delta V = 0.7\,\text{C}\times6.7\,J$

$\Delta V = 4.69\,\text{V}$

$V = 1000\,\text{V}\,\text{m}^{-1}\times0.1\,\text{m}$

$V = 100\,\text{V}$

$E = qV$

$E = 1.6\times10^{-19}\,\text{C}\times100\,\text{V}$

$E = 1.6\times10^{-17}\,\text{J}$

$E = \frac{1}{2}mv^2$

$v = \sqrt{\frac{2E}{m}}$

$v = \sqrt{\frac{2\times1.6\times10^{-17}}{9.11\times10^{-31}}}$

$v = 5.93\times10^{5}\,\text{ms}^{-1}$

$\therefore \Delta U = 1.6\times10^{-15}\,\text{J}$

$KE = \frac{1}{2}mv^2$

$v = \sqrt{\frac{2U}{m}}$

$v = \sqrt{\frac{2\times1.6\times10^{-15}}{9.11\times10^{-31}}}$

$v = \sqrt{3.51\times10^{15}}$

$v = 5.93\times10^{7}\,\text{ms}^{-1}$

This is the energy the electron has just before it hits the target. Now we can find the velocity from $KE = \frac{1}{2}mv^2$ as follows:

$$U = \frac{1}{2}mv^2$$

$$v = \sqrt{\frac{2U}{m}}$$

$$v = \sqrt{\frac{2 \times 1.6 \times 10^{-15}}{9.11 \times 10^{-31}}}$$

$$v = \sqrt{3.51 \times 10^{15}}$$

$$v = 5.93 \times 10^{7}\,\text{m s}^{-1}$$

QUESTIONS

1 Define electric potential energy.

---

2 Can electric potential be negative? Explain.

---

3 What is the potential difference between two points A and B in an electric field if the work done on a charge of 0.7 C is 6.7 J when moving it from A to B? State which point has the higher potential (A or B).

---

4 Two parallel plates have opposite charges, positive and negative and are separated by a distance of 10 cm. The potential difference between the parallel plates is 1000 V m^{-1}. Calculate the potential difference between the plates and find the velocity an electron would have when moving from the negative plate to the positive plate when it arrives at the positive plate.

9780170412643

5 The zero of potential energy can be defined as the negative plate in the previous question. With this in mind, what would be the potential of a charge q when it is midway between the parallel plates? State whether the charge's potential would increase or decrease when the charge was left to move freely in this electric field.

EVALUATION

Multiple-choice

1 When the distance between two identical point charges is reduced by a quarter, what happens to the force between them?

A The force reduces by a quarter.

B The force increases by a quarter.

C The force reduces by more than a quarter.

D The force increases by more than a quarter.

2 An electric field is not affected by:

A charges travelling in the field.

B charges exerting the field.

C the distance away from the source of the field.

D the medium the field is in.

3 Potential difference is the same as:

A potential energy.

B kinetic energy.

C voltage.

D work done.

Short answer

4 Two charges $q = 5\,\mu C$ and $Q = 10\,\mu C$ are separated by distance 0.05 m. Determine the magnitude of force exerted on q by Q.

9780170412643

5 Two parallel plates are situated vertically, where the positive plate is on the bottom and the negative plate is suspended some distance d above it. A small positive electric charge of charge 2.0×10^{-6} C and mass 3.6×10^{-6} kg is suspended in the middle of this field, motionless. Calculate the size of the electric field.

6 The table below shows how electric field strength changing at increasing distances from the source.

r (m)	E (N C^{-1})
0.02	17.5
0.04	4.41
0.06	1.97
0.08	1.10
0.10	0.72

Construct a graph by first making the data linear to determine the strength of the point source emitting this field.

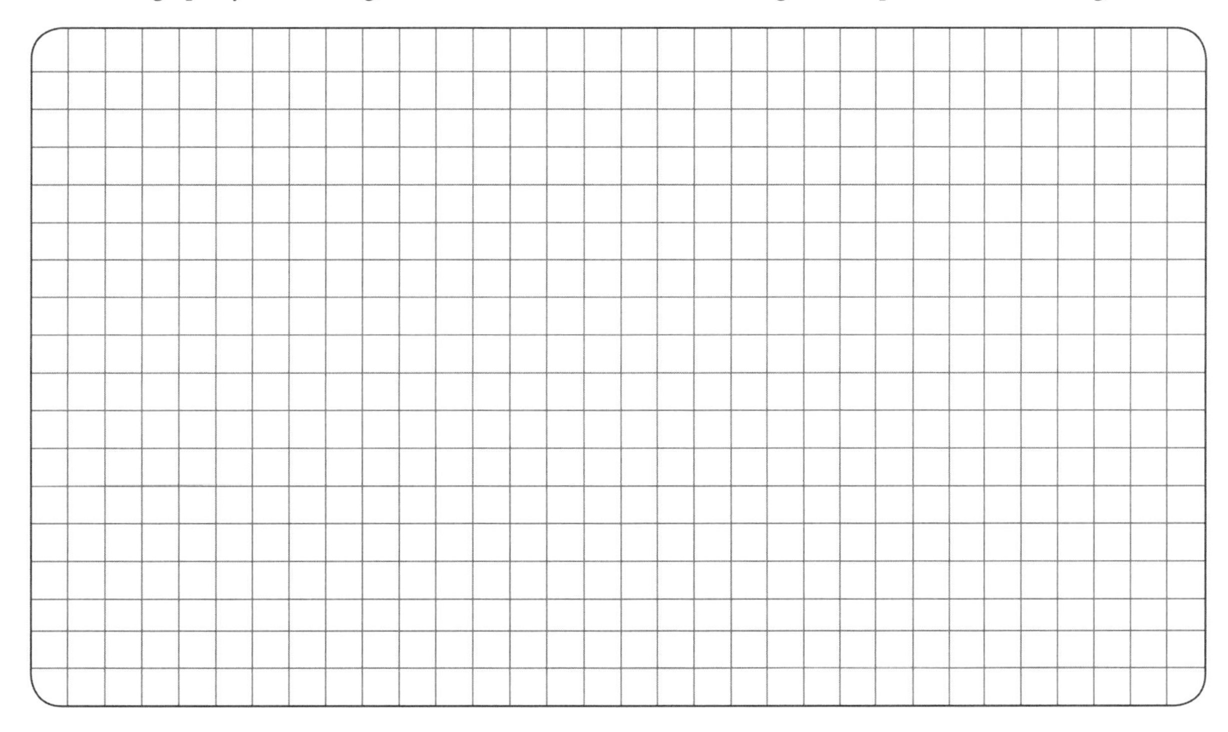

8 Magnetism

Summary

- Magnets have a north pole and south pole.
- Metals in Earth's crust with intrinsic magnetic properties can be magnetised to make permanent magnets.
  - These intrinsic properties include a half full outer shell of electrons and have all magnetic domains aligned.
- Magnetic field lines are drawn out of the north pole and into the south pole.
- Materials can be classified as diamagnetic, paramagnetic and ferromagnetic.
  - Diamagnetic materials are weakly repelled by nearby magnets.
  - Paramagnetic materials are weakly attracted to nearby magnets.
  - Ferromagnetic materials are strongly attracted to nearby magnets.
- Magnetic field strength, B, is measured in Tesla's and is a very large unit.
- Electric current induces a magnetic field in concentric circles about the wire and the field strength be modelled as $B = \frac{\mu_0 I}{2\pi r}$ where B is the magnetic field strength in T, μ_0 is the permittivity of free space ($4\pi \times 10^{-7}$ Tm A^{-1}), I is the current through the wire in A and r is the distance in m away from the wire.
  - The direction of the circular B field about a wire is modelled with Maxwell's screw rule. By giving a *thumbs up* with your right hand and pointing your thumb in the direction of conventional current, your fingers curl in the direction of the B field about the wire.
- A solenoid is a coil of wire with a hollow core, that when electric current is put through turns into an electromagnetic where the field strength varies with a number of factors. Modelled by: $B = \mu_0 nI$ where n is the number of turns per meter in the solenoid, and B is the magnetic field strength within the core of the solenoid.
  - The direction of the north pole in a solenoid can be modelled with an adaptation of Maxwell's screw rule. If you use your right hand to give a *thumbs up* and your fingers are curling in the direction of the current flowing through the solenoid, then your thumb points in the direction of the north pole of the solenoid.
- When charge is in an external B field, a force will act on that charge.
  - When the charge is moving in a wire with current I, the force exerted on the wire is modelled as $F = BIL\sin\theta$.
  - When a charge q is moving on its own in an external B field, it will experience a force modelled by $F = qvB\sin\theta$.

- Charges and currents moving perpendicular to an external B field will experience a maximum force, where charges and currents parallel to external B fields will experience no force.
- The direction of force on a moving charge or current is determined by the right-hand rule, where the thumb points in the direction of the moving charge/current, fingers point in the direction of the external B field, and the palm pushes in the direction of the force acting on the charge/wire.

8.1 Magnetic properties

QUESTIONS

1 Compare diamagnetic, paramagnetic and ferromagnetic materials.

2 Draw the magnetic field about a bar magnetic.

3 How could the strength of a bar magnetic be represented with field lines?

4 If a current is moving in a wire coming out of the page, draw the magnetic field lines about this wire.

9780170412643

5 Explain why iron filings align with the magnetic field lines about permanent magnets.

8.2 Magnetic fields due to moving charge

WORKED EXAMPLES

1 Calculate the magnetic field strength at a perpendicular distance of 15 cm from a long wire carrying a current of 5 A.

2 If this wire was then curled into a solenoid with 20 turns over 5 cm, what would be the magnetic field strength in the core of the solenoid?

ANSWERS

1 $B = \frac{\mu_0 I}{2\pi r}$

$B = \frac{4\pi \times 10^{-7} \times 5\,\text{A}}{2\pi \times 0.15}$

$B = 6.67 \times 10^{-6}\,\text{T}$

2 $B = \mu_0 n I$

$B = 4\pi \times 10^{-7} \times \frac{20}{5} \times 5$

$B = 2.51 \times 10^{-5}\,\text{T}$

QUESTIONS

1 How far away from a wire carrying a current of 15 A will the B field have a magnitude of 7.0×10^{-4} T?

2 Calculate the current needed to flow through a wire to produce a B field of 8.9×10^{-5} T at 10 cm from the wire.

3 If a solenoid of 200 turns per meter has a current of 6 A running through it, what is the magnitude of the B field within the core?

4 A solenoid carrying 15 A of current has a B field of 6.6×10^{-2} T. Determine the number of turns in this solenoid.

5 In the centre of a solenoid, is the B field uniform or non-uniform? Explain.

6 Explain the two possible outcomes when a bar magnetic is brought close to the hollow core of an electromagnet.

7 If it takes nine field lines to show the B field about a ring of current carrying wire, how many field lines would be required to show the field if 100 of these rings were packed closely together to simulate an electromagnet?

8.3 Forces on charges and current carrying conductors in magnetic fields

WORKED EXAMPLES

1 A wire carrying current of 2 A is placed perpendicular to a B field of strength 7.4×10^{-4} T as shown below. If only 1.5 m of wire is in the B field, what is the magnitude and direction of the force acting on the wire?

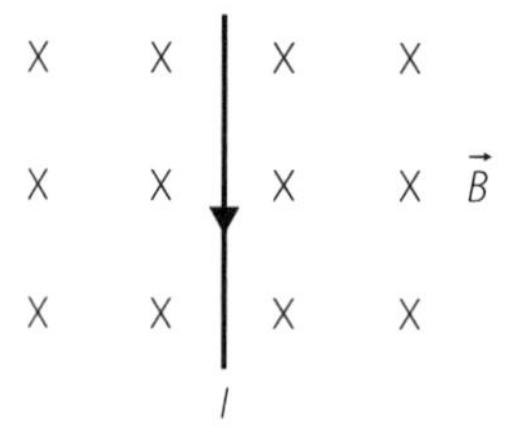

X X X X
X X X X $\vec{B}$
X X X X
I

2 An electron is ejected into a B field with velocity 5.0×10^{5} ms^{-1} as shown below. It experiences a force of 5×10^{-10} N downwards when entering the field. What is the magnitude of the B field?

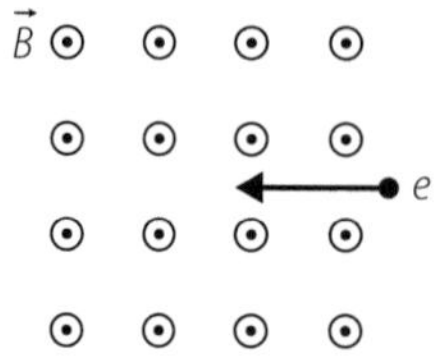

$\vec{B}$
e

ANSWERS

1 Using the right-hand rule, it can be determined that the palm will point to the right when the thumb points down and the fingers go into the page, therefore the wire will move to the right.

The magnitude of the force can be calculated as follows:

$$F = BIL\sin\theta$$

$$F = 5.0\times10^{-2}\,\text{T}\times3.0\,\text{A}\times0.5\,\text{m}\times\sin60$$

$$F = 6.50\times10^{-2}\,\text{N}$$

$$F = BIL\sin\theta$$

$$B = \frac{F}{IL\sin\theta}$$

$$B = \frac{4\,\text{N}}{8.0\,\text{A}\times0.6\,\text{m}\sin0}$$

$$B = 0.83\,\text{T}$$

2

$$F = qvB\sin\theta$$

$$F = 2\times1.6\times10^{-19}\,\text{C}\times4.8\times10^{4}\,\text{ms}^{-1}\times1.9\times10^{-2}\,\text{T}\sin0$$

$$F = 2.9\times10^{-16}\,\text{N}$$

$$F = qvB\sin\theta$$

$$B = \frac{F}{qv\sin\theta}$$

$$B = \frac{2.5\times10^{-15}\,\text{N}}{1.6\times10^{-19}\,\text{C}\times v\times\sin45}$$

$$B = \left(\frac{22097}{v}\right)T$$

QUESTIONS

1 Explain how the path of a proton and electron differ when entering the same B field.

____

____

____

2 A wire carrying a current of 3.0 A is placed in a uniform B field of strength 5.0×10^{-2} T. Find the force on the wire if only 0.5 m of the wire is in the B field, and the wire and the B field have an angle of 60° between them.

____

____

____

3 A wire carrying a current of 8.0 A and 0.6 m is placed perpendicular to a uniform magnetic field and experiences a force of 4 N. What is the size of the magnetic field?

____

____

____

4 What is the minimum magnitude of a magnetic field necessary to apply a force of 9.9×10^{-11} N to an electron moving at a speed of $200\,\text{km}\,\text{s}^{-1}$?

5 When a moving charge enters a magnetic field, it experiences a force. Consider the figure from the second worked example. Describe how the motion of the electron would change over time as it continues to travel through the *B* field. Draw a diagram to help you.

Multiple-choice

1 A magnetic field exists about a current carrying wire. At 5 cm from the wire, the field strength is 3.0×10^{-6} T. The field is 6.0×10^{-6} T at:

A 10 cm from the wire.

B 6 cm from the wire.

C 5 cm from the wire.

D 3.5 cm from the wire.

2 An electron enters a B field into the page from the negative x direction. The direction of force acting on the electron is:

A into the page.

B out of the page.

C up.

D down.

3 Electromagnets have the potential to have very large magnetic fields. Why?

A They are stronger than permanent magnets.

B The current can be adjusted which adjusts the magnetic field in the core respectively.

C Electrons move more quickly in coils than in straight wires.

D Being in a coil reduces the resistance, and increases the current, hence increases the B field.

Short answer

4 An α particle enters a magnetic field from the right with velocity 4.8×10^4 m s^{-1}. The B field is coming out of the page and has a magnitude of 1.9×10^{-2} T. What is the magnitude and direction of force acting on the α particle?

5 An electron enters a magnetic field from the left at 45° to the field and experiences a force of 2.5×10^{-15} N downwards. Calculate the magnitude and direction of the B field in terms of v.

6 How does the B field in a solenoid change if the length of the solenoid doubles and the number of turns remains the same?

9780170412643

9 Electromagnetic induction

Summary

- Electromagnetic induction is the production of an electric field by a changing magnetic field.
- The induced electric field creates an electromotive force (emf) that may generate a current in a conductor.
- The magnetic flux (Φ) through a surface, with units of Webbers (Wb) is a measure of the total magnetic field passing through a surface. It is calculated as: $\Phi = BA\cos\theta$, where B is the magnitude of the magnetic field, A is the area of the surface and θ is the angle between the magnetic field lines and normal to the surface.
- Faraday's law of induction states that the induced emf, with units of volts (V), in a loop of wire is equal to the negative of the time rate of change of the magnetic flux passing through it: $\text{emf} = -\frac{\Delta\Phi}{\Delta t}$, where $\Delta\Phi$ is the change in magnetic flux through the surface, and Δt is the time interval over which the change takes place.
- Lenz's law states that the an induced emf generates a current that travels in a direction that causes a magnetic flux change that opposes the original change in magnetic flux that induced the emf.
- An alternating current (AC) generator consists of a rotating armature, composed of multiple loops of wire, within a magnetic field.
- Current induced by an AC generator follows a sinusoidal pattern that we refer to as alternating current.
- The root mean square current or voltage is the average AC current or voltage that produces the same power as a DC current or voltage of the same magnitude: $V_{RMS} = \frac{V_{peak}}{\sqrt{2}}$, $I_{RMS} = \frac{I_{peak}}{\sqrt{2}}$ and $P_{ave} = V_{RMS} \times I_{RMS}$.
- A transformer uses electromagnetic induction to change the voltage and current of a primary AC.
- A transformer consists of a primary and a secondary coil of wire wrapped around an iron core. The changing magnetic flux produced in the primary coil is transmitted through the iron core to the secondary coil, which in turn produces an emf through electromagnetic induction.
- The magnitude of the secondary voltage or current can be calculated as: $\frac{V_S}{V_P} = \frac{I_P}{I_S} = \frac{N_S}{N_P}$, where N indicates the number of coils in the primary (P) or secondary (S) coils.
- A step-up transformer has $N_P < N_S$; therefore, $V_P < V_S$. A step-down transformer has $N_P > N_S$; therefore, $V_P > V_S$

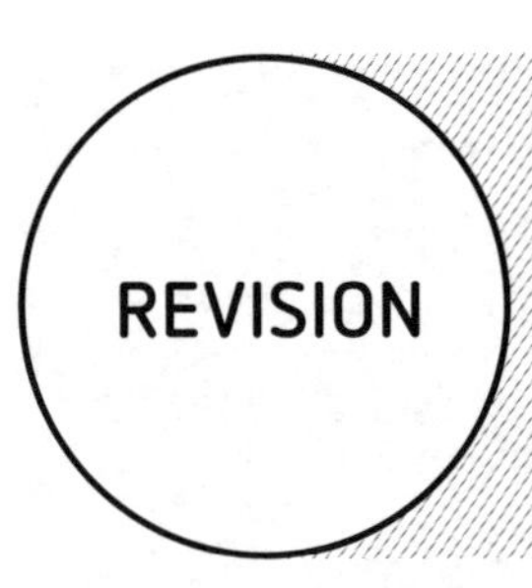

9.1 Electromagnetic induction

Michael Faraday discovered that a change in the amount of a magnetic field through a surface, a feature that he termed the magnetic flux, Φ, could produce an electric field. The magnetic flux measured in Webbers (Wb) is calculated as $\Phi = BA\cos\theta$. A changing flux in a region induces an electromotive force (emf) in that region that will produce an electric current if it acts upon free charge carriers.

QUESTIONS

1 Define the following terms.

a Electromagnetic induction

____

____

b Electromotive force

____

____

c Induced current

____

____

d Magnetic field

____

____

e Magnetic flux

____

____

f Magnetic flux density

____

____

2 Determine the magnetic flux passing through a square surface of side length 22 cm if it is orientated perpendicularly to a uniform 0.10 T magnetic field.

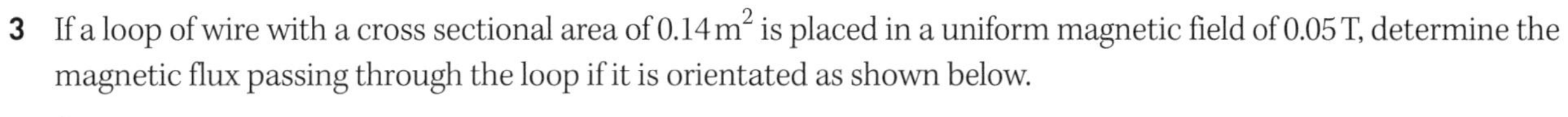

3 If a loop of wire with a cross sectional area of $0.14\,m^2$ is placed in a uniform magnetic field of 0.05 T, determine the magnetic flux passing through the loop if it is orientated as shown below.

a

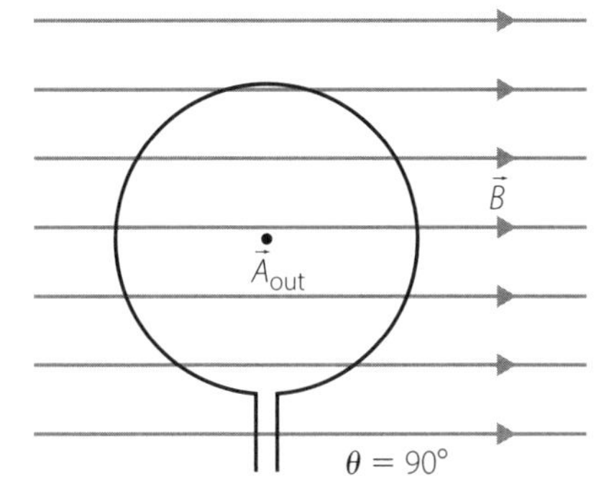

b

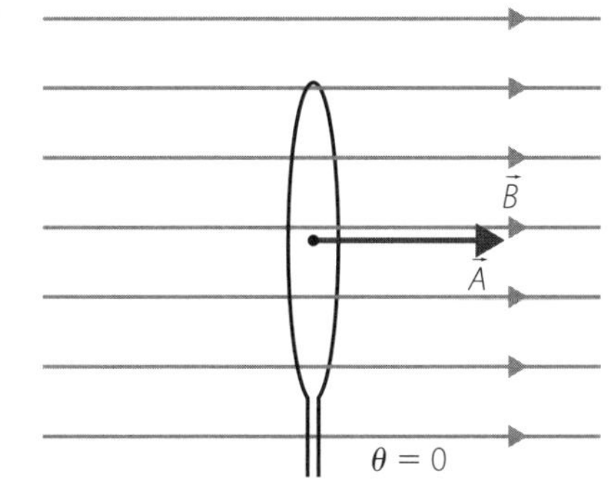

4 Determine the angle of orientation between the normal to a circular loop of wire with a radius of 5.0 cm and a uniform 1.4 T magnetic field if it is known that a magnetic flux of 9.5 mWb passes through the loop.

9.2 Faraday's law of induction

Faraday's law of induction states that a time changing magnetic flux passing through a conducting loop of wire will induce an electromotive force with a magnitude equal to the negative of the rate of change of flux: $\text{emf} = -\frac{\Delta\Phi}{\Delta t} = -\frac{\Delta\Phi_f - \Delta\Phi_i}{\Delta t}$. The emf has units of volts (V). This emf will generate a current in the loop with a magnitude that can be determined by Ohm's law: $I = \frac{\text{emf}}{R}$, where R is the resistance of the circuit.

QUESTIONS

1 Insert the following phrases into the provided flow diagram to illustrate the generation of an electric current in a loop of wire due to the phenomenon of electromagnetic induction.

Electromotive force	Final magnetic flux
Initial magnetic flux	Induced current

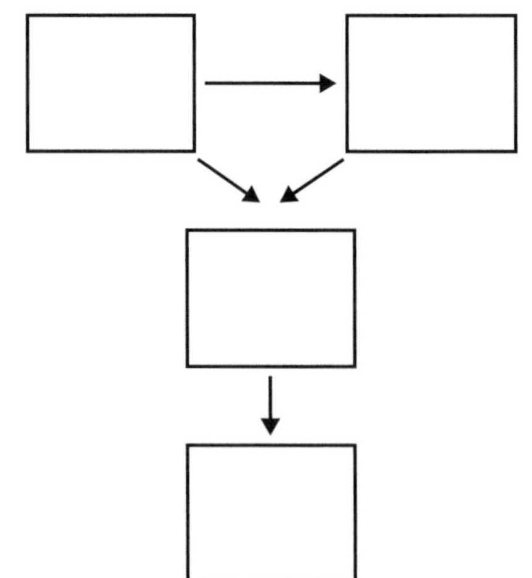

2 If a square loop of wire is placed perpendicularly in a large uniform magnetic field, determine whether the following changes made would result in the induction of a current in the wire. Justify your answers.

a The loop of wire is rotated.

9780170412643

b The magnetic field strength is increased.

c The loop is moved in a straight line within the magnetic field.

d The size of the loop is decreased.

9.3 Solving problems using Faraday's law

By inspecting Faraday's law, it can be seen that an emf can be produced in one of three ways: changing the magnetic field density (B), changing the area of the coil (A), or by changing the angle (θ) between the magnetic field and the area. If there are multiple loops in the coil, Faraday's law can be rewritten as: emf $= \frac{n\Delta\Phi}{\Delta t}$, where n is the number of coils.

QUESTIONS

1 If a single loop of wire with a radius of 0.060 m is placed perpendicularly within a uniform 0.050 T magnetic field, determine the emf produced if it is completely removed from the magnetic field in 0.050 s.

2 If a coil of conducting wire containing 10 loops of cross sectional area 10.0 cm^2 is placed perpendicularly within a uniform 1.0 T magnetic field, determine the induced current within the loop if it has a resistance of 0.50 Ω and is rotated by 180° in 0.015 s.

3 A coil of wire consisting of 100 loops is placed in a uniform magnetic field. If the magnetic flux through a single loop of wire changes as shown on the graph below, determine the current induced in the loop of wire if it has a resistance of 2.0 Ω.

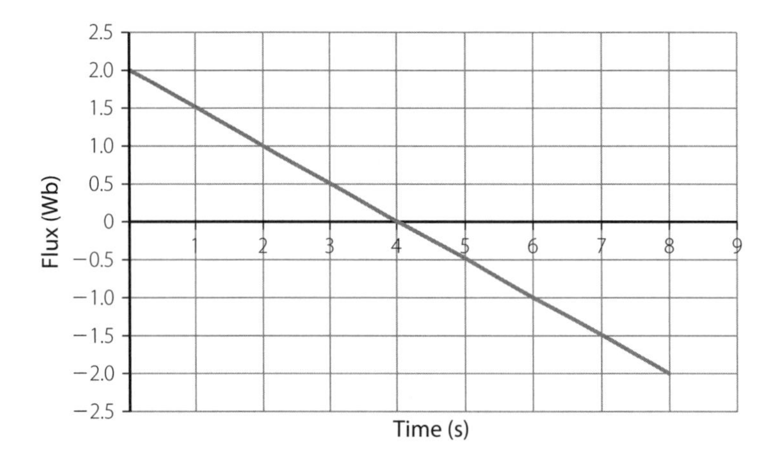

9.4 Lenz's law

Lenz's law uses the conservation of energy as a means to predict the direction of the current induced by a time changing magnetic flux. It states that the induced current flows in a direction such that the magnetic flux produced by the current is orientated to oppose the change in magnetic flux that originally produced it. If this was not the case, the change in magnetic flux; and therefore, the induced current, would continue to grow and produce more energy than was lost by the magnetic flux.

QUESTIONS

1 Insert the phrases provided in column A into the boxes, and the phrases in column B onto the arrows of the blank flow chart provided below; which describes the production of a current in a conducting loop of wire due to a changing magnetic flux.

COLUMN A	COLUMN B
An induced current around the surface	That produces
A changing magnetic flux through a conducting surface	That generates
A magnetic field through the surface	Which opposes the
An electromotive force in the surface	Induces

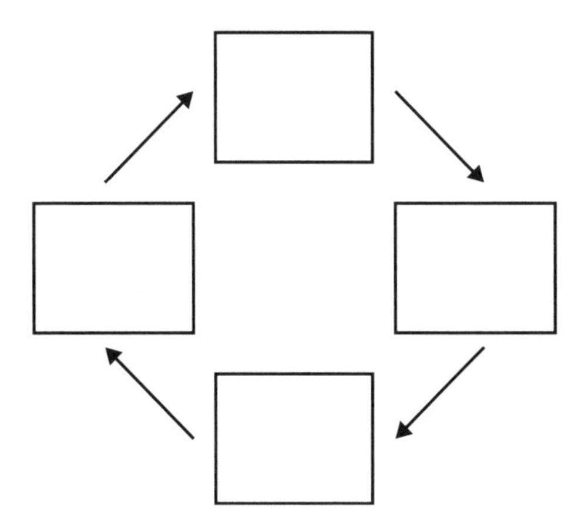

2 In each of the following examples, determine whether the induced current in the conducting loop of wire will travel clockwise or anti-clockwise.

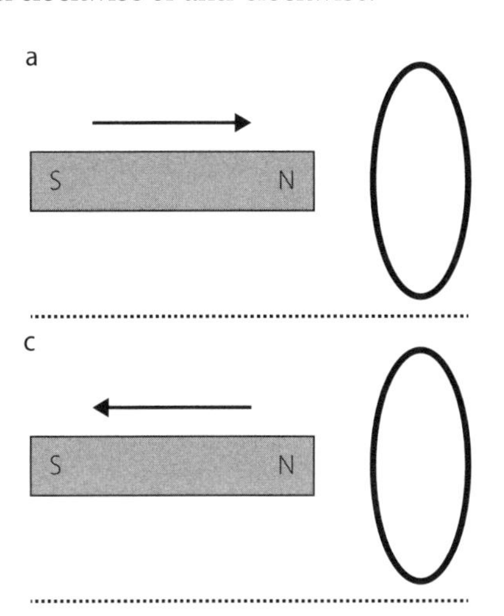

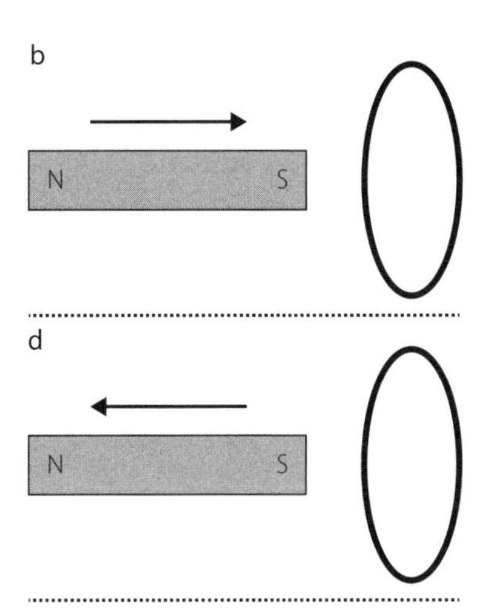

9.5 Production and transmission of alternating current

An alternating current generator consists of an armature containing multiple of loops of conducting wire connected to an external circuit. When the armature is placed within a magnetic field and is uniformly rotated relative to that magnetic field, a current which varies sinusoidally with time is induced. This current is termed alternating current (AC). The root mean square (RMS) voltage and current of AC is the average current or voltage that produces the same power as a DC current or voltage of the same magnitude: $V_{RMS} = \frac{V_{peak}}{\sqrt{2}}$, $I_{RMS} = \frac{I_{peak}}{\sqrt{2}}$ and $P_{ave} = V_{RMS} \times I_{RMS}$.

A transformer, which consists of two coils of wire wrapped around an iron core, is able to change the voltage and current of AC by the formula $\frac{V_S}{V_P} = \frac{I_P}{I_S} = \frac{N_S}{N_P}$, where N indicates the number of coils in the primary (P) or secondary (S) coils. A step-up transformer has $N_P < N_S$; therefore, $V_P < V_S$. A step-down transformer has $N_P > N_S$; therefore, $V_P > V_S$.

QUESTIONS

1 Label the diagram of an AC generator below with the terms listed below.

Axis of rotation	Alternating emf output
emf	Armature
Slip rings	

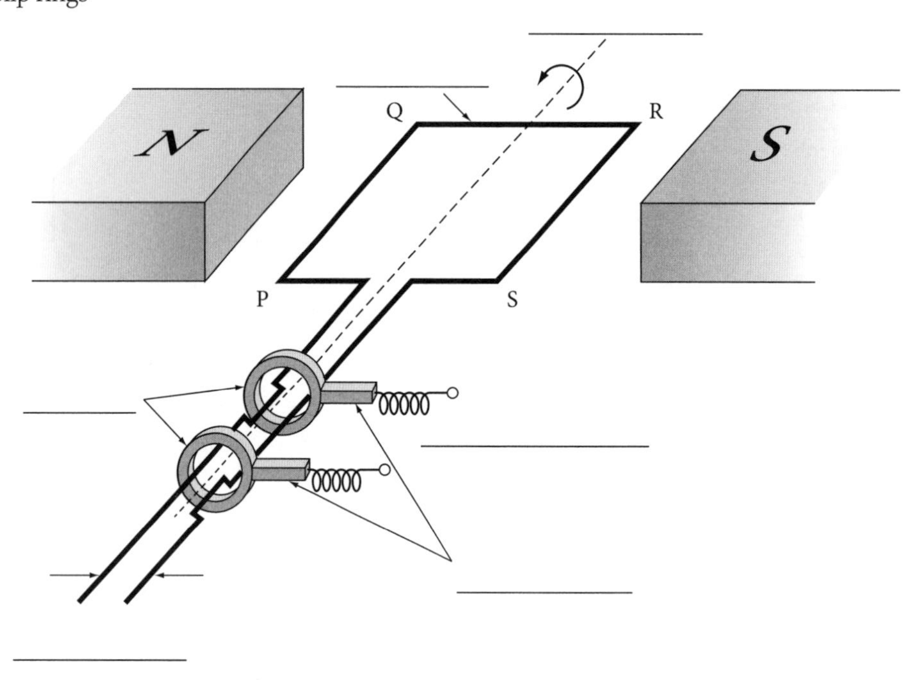

2 Use the graph below to answer the following questions.

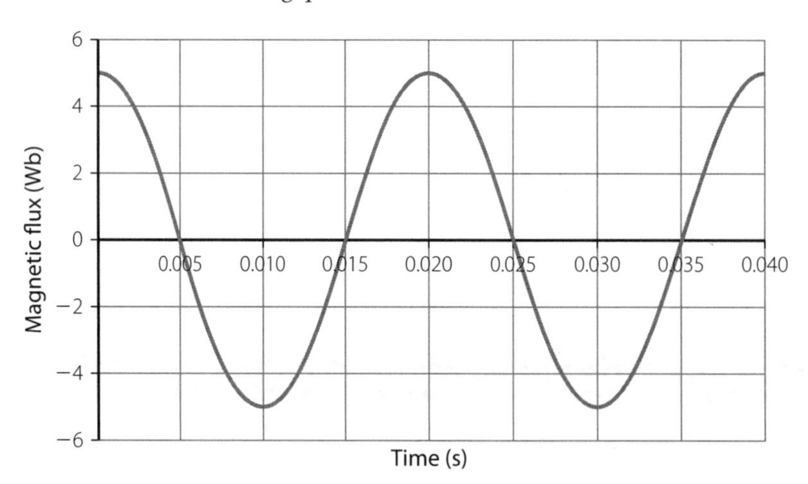

a Determine an equation for the magnetic flux (Φ) as a function of time (t).

9780170412643

b Determine an equation for the emf as a function of time.

c Calculate the peak voltage and root mean square voltage.

d Calculate the peak current and root mean square current if the armature has a resistance of 2.5 Ω.

e Calculate the average power of the generator.

3 A 120 W, 12 V AC supply is connected to the input terminals of a transformer. The primary coil is wound with 1500 turns. The output emf is 60.0 V. Assume there is no power loss in the transformer.

a Find the number of turns on the secondary coil.

b Is this a step-up or step-down transformer?

c What is the output current?

Multiple-choice

1 When an electric field is induced in a conductor, which of the following options is said to exist?

A Electromotive force

B Electrolysis

C Electromagnetic induction

D Electricity

2 Which of the following would result in an increase in the magnetic flux passing through a substance?

A A decrease in the strength of the magnetic field.

B A reduction in the area of the surface.

C The surface rotates to become more perpendicular to the magnetic field.

D The surface rotates to become more parallel to the magnetic field.

3 The electromotive force has units of Volts, what is another way that this unit could be written?

A $\mathrm{m\,T}$

B $\mathrm{T\,m^{-2}}$

C $\mathrm{T\,m^{2}}$

D $\mathrm{T\,m^{2}\,s^{-1}}$

4 If a coil of wire, initially orientated perpendicularly to a magnetic field, is rotated about its axis, a graph of the emf produced would produce:

A a linear curve.

B a cosine curve.

C a sine curve.

D an exponential curve.

5 Which of the following is true for a step-up transformer?

A $V_P > V_S$

B $I_S > I_P$

C $N_P < N_S$

D $P_P < P_S$

9780170412643

Short answer

6 What is the term given to the number of magnetic field lines passing through a given area?

____

7 In what direction relative to the changing magnetic flux, does the magnetic field generated by an induced electric current face?

____

8 What is the term given to the average AC potential difference that produces the same power in a load as a DC potential difference of the same magnitude?

____

9 Explain the process of inducing an electric current in a conductor by changing the strength of the magnetic field.

____

____

____

____

____

10 Explain how Lenz's law agrees with the conservation of energy.

____

____

____

____

____

11 If a square solenoid consisting of 120 loops of wire of side length 10.0 cm is initially orientated perpendicularly to uniform magnetic field of 0.15 T, calculate the emf that would be produced if it is rotated until a normal to its surface is at an angle of 25° to the magnetic field lines over a period of 0.20 s.

____

____

____

____

____

____

____

12 If the radius of a conducting loop of wire is increased from 10.0 cm to 25 cm over a period of 1.5 s while it is in a uniform magnetic field of 1.0 T, calculate the magnitude of the emf produced if the magnetic field and the loop are perpendicular to each other.

13 If an AC generator consists of 250 loops with an area of 0.050 m^2 rotating in a magnetic field of 0.75 T has a period of rotation of 0.20 s, determine the following.

a An equation describing the behaviour of the magnetic flux through the loops as a function of time.

b An equation describing the behaviour of the induced emf as a function of time.

c The peak voltage produced.

d The root means square of the voltage produced.

e The average power produced if the armature has a resistance of 0.05 Ω.

9780170412643

10 Electromagnetic radiation

LEARNING

Summary

- Electromagnetic waves are produced by oscillating electric charges.
- Electromagnetic waves consist of self-sustaining and mutually perpendicular oscillations in the magnetic and electric fields that travel as transverse waves.
- Electromagnetic waves do *not* require a medium to travel through as they consist of electric and magnetic field oscillations.
- The speed of light in a vacuum (c) is equal to $3.0 \times 10^8\,\mathrm{m\,s^{-1}}$.
- The speed of light in any medium can be calculated as the inverse of the square root of the product of the electrical permittivity (ε) and the magnetic permeability (μ) of the medium. $v = \frac{1}{\sqrt{\varepsilon_0 \mu_0}}$
- The velocity of the wave is equal to the frequency of field oscillation (f) multiplied by the wavelength of the wave (λ).
- Visible light forms only one part of the electromagnetic spectrum which contains all of the possible electromagnetic waves arranged by wavelength, frequency and energy.
- Transmitting antennas use AC as the initial source of oscillating electric fields that produce electromagnetic waves with the same frequency as the AC.
- When electromagnetic waves interact with a receiving antenna, the oscillating electromagnetic fields produce an AC that oscillates at the same frequency as the electromagnetic wave.

10.1 Electromagnetic waves

Maxwell's equations described electromagnetic waves as oscillations in the electric and magnetic fields. They could describe the velocity of electromagnetic waves, the fact that they could travel in the absence of a medium, the wave behaviour of light and predicted the existence of a range of electromagnetic waves outside of the visible spectrum. The initial source of these oscillations is generally an oscillating electric charge, a fact that is utilised by transmitting antennas.

QUESTIONS

1 Draw an image of the oscillations in the electric and magnetic fields of an electromagnetic wave travelling to the right of the page,

2 Determine the magnetic permeability of a vacuum if the electrical permittivity is measured to be $8.85 \times 10^{-12}\,\mathrm{F\,m^{-1}}$.

3 An electromagnetic wave of frequency (f) 1.5×10^{18} Hz is travelling in a vacuum. Determine the:

a region of the electromagnetic spectrum it is travelling in

b wavelength of the electromagnetic wave.

9780170412643

Multiple-choice

1 Which of the following describes the orientation of the direction of travel, the oscillations in the electrical field and the oscillations in the magnetic field of an electromagnetic wave?

A Parallel

B Perpendicular

C Collinear

D Congruent

2 In which region of the electromagnetic spectrum does a wave with a frequency of 1.0×10^6 Hz belong?

A Radio waves

B Microwaves

C Visible light

D X-rays

3 Which of the following has the shortest wavelength?

A Radio waves

B Microwaves

C Visible light

D X-rays

4 Which of the following would *not* give rise to an electromagnetic wave?

A Alternating current

B Direct current

C An accelerating electron

D An accelerating proton

Short answer

5 What type of wave is an electromagnetic wave?

____

6 What is the term given to a diagram of all possible electromagnetic waves ordered by increasing frequency?

____

7 Rearrange the following regions of the electromagnetic spectrum in order of decreasing wavelength: UV rays, gamma rays, visible light, radio waves, microwaves, infrared radiation and X-rays.

____

____

8 Explain how an electromagnetic wave is created in a transmitting antenna.

____

____

____

____

____

9 Explain how an electromagnetic wave is received by a receiving antenna.

____

____

____

____

____

10 An electromagnetic wave of frequency 3.0×10^{10} Hz travels through a medium that has an electrical permittivity of $5.6 \times 10^{-10}\,\mathrm{F\,m^{-1}}$ and a magnetic permeability of $1.3 \times 10^{-6}\,\mathrm{T\,m\,A^{-1}}$, determine the:

a region of the electromagnetic spectrum that the wave lies in

____

b velocity of the wave

____

____

____

c wavelength of the wave.

____

____

____

9780170412643

UNIT FOUR

REVOLUTIONS IN MODERN PHYSICS

11 Special relativity

Summary

- Special relativity describes what happens to objects with mass when moving close to the speed of light.
- A frame of reference can be accelerating or moving with constant velocity and describes motion according to a coordinate system.
- An inertial frame of reference is one where Newton's first law applies to a very good approximation. This is a non-accelerating frame of reference.
- Motion can only be measured relative to an observer.
- There are two postulates of special relativity.
  - The laws of physics are the same in all inertial frames of reference – the principle of special relativity.
  - The speed of light has the same value, c, in all inertial frames. It does *not* depend on the speed of either the source or the observer.
- The concept of simultaneity suggests that when two events occur simultaneously in one reference frame they *cannot* occur simultaneously in another reference frame, if the other reference frame is moving relative to the first reference frame.
- When objects travel close to the speed of light, they experience time dilation, length contract and relativistic momentum.
  - Time dilation is a longer time measured from an observer outside the reference frame of an event. Modelled by $t = \frac{t_0}{\sqrt{1-\frac{v^2}{c^2}}}$ where t_0 is the proper time (s) measured in the same reference frame as the event, t is the dilated time (s), v is the relative velocity between the inertial reference frames ($m\,s^{-1}$) and c is the speed of light ($3 \times 10^8\ m\,s^{-1}$).
  - Length contraction is a shorter length measured in a reference frame that is moving relative to an inertial frame. Modelled by $L = L_0\sqrt{1-\frac{v^2}{c^2}}$ where L_0 is the proper length (m) measured when the object is at rest, and L is the contracted length (m).
  - Relativistic momentum is the momentum of a particle due to the relativistic mass at very high speeds. Modelled by $p_v = \frac{p_0}{\sqrt{1-\frac{v^2}{c^2}}}$ where p_0 is equal to the product of the rest mass (kg) and the relative velocity v ($m\,s^{-1}$), and p_v is the relativistic momentum (Ns)

9780170412643

- The Lorentz factor $\gamma = \frac{1}{\sqrt{1-\frac{v^2}{c^2}}}$ is used to simplify writing equations for time dilation, length contraction and relativistic momentum.
- The mass energy equivalency states $E = mc^2$ where E is the energy change (J) equivalent to the change in mass (kg).
- Objects with mass can never travel at the speed of light because they become heavier as they travel faster. An infinite amount of energy would be required to move a mass this heavy at this speed: an impossible reality.

REVISION

11.1 Relative motion

QUESTIONS

1 A person is walking north at $2\,m\,s^{-1}$ and is passed by a car moving at $20\,m\,s^{-1}$. Calculate the relative motion of the car with respect to the person, and the person with respect to the car.

2 A train travelling east at $25\,m\,s^{-1}$ meets a train travelling west at $20\,m\,s^{-1}$. According to a passenger on the westbound train, what is the velocity of the eastbound train?

3 A boat moving is $15\,m\,s^{-1}$ north parallel to a nearby shore. On board, a child throws a ball at $3\,m\,s^{-1}$ north.

a For a sunbather on the beach, what speed does the ball travel north?

b What speed would an on-board passenger measure the ball at?

11.2 Definitions and simultaneity

QUESTIONS

1 A boat is approaching a lighthouse in the evening. The lighthouse keeper sends out a beacon of light. If the boat is moving at speed 0.5c what speed does the captain on the boat observe the lighthouse beacon?

2 As a rocket ship moves past earth at speed v and sends out a light pulse, what speed do observers on earth see the light travelling at?

3 Paddy stands on the platform while Sian rides by at 0.5c on her cart. At the instant that she and Paddy are next to each other, two fire-crackers go off at either end of Sian's cart. In Sian's frame of reference, the two flashes of light, from equidistant points, reach her at the same instant – so the ignition of the two firecrackers were events. According to Paddy, which flash – from the front or the rear of her cart – will reach Sian first? Justify your answer.

4 You are standing between two trees A and B that are 1 km apart during a storm. Your friend stands at the base of tree A trying to foolishly seek shelter. Lightning strikes both trees simultaneously from the perspective of your friend. Explain when you see the lightning strike the trees.

11.3 Time dilation, length contraction and relativistic momentum

WORKED EXAMPLE

A spaceship leaves Earth with velocity 0.85c to travel to a star 5 ly away. How long will the trip take according to:

1 Earth's clocks?

2 the spaceship's clocks?

3 What distance does the spaceship travel according to observers on Earth?

ANSWERS

1 According to observers on Earth, the distance to the star is 5 ly, and the spaceship travels at 0.85c.

$1\,\text{ly} = 3 \times 10^8\,\text{m s}^{-1} \times 3.16 \times 10^7\,\text{s}$

$1\,\text{ly} = 9.47 \times 10^{15}\,\text{m}$

So:

$$t = \frac{d}{v}$$

$$t = \frac{5 \times 9.47 \times 10^{15}\,\text{m}}{0.85 \times 3 \times 10^8\,\text{m s}^{-1}}$$

$$t = 1.86 \times 10^8\,\text{s}$$

2 The time measured on the spaceship is the proper time t_0. This clock runs slower than the clock on Earth.

$$t = \frac{t_0}{\sqrt{1 - \frac{v^2}{c^2}}}$$

$$t_0 = t\sqrt{1 - \frac{v^2}{c^2}}$$

$$t_0 = 1.86 \times 10^8\,\text{s}\sqrt{1 - \frac{(0.85c)^2}{c^2}}$$

$$t_0 = 9.80 \times 10^7\,\text{s}$$

3 The spaceship will measure a contracted length. The proper length is measured from Earth as 5 ly (4.74×10^{16} m).

$$L = L_0\sqrt{1 - \frac{v^2}{c^2}}$$

$$L = 4.74 \times 10^{16}\,\text{m}\sqrt{1 - \frac{(0.85c)^2}{c^2}}$$

$$L = 2.50 \times 10^{16}\,\text{m}$$

QUESTIONS

1 Satellites used for the Global Positioning System travel at 14 000 km h^{-1} relative to Earth. In which frame, on Earth or on the satellite, would proper time (one day) be measured?

____

____

____

9780170412643

2 Amy flies from Perth to Brisbane to visit her friend Tara. Amy measures the distance the plane travels as she knows the velocity v of the plane and how long, t, it takes to get to Brisbane. Tara asks her employees to measure the ground from Perth to Brisbane with a trundle wheel. Who, if anyone, measures the shorter distance from Perth to Brisbane?

3 Eleni watches Mario, travelling at 0.8c to the left relative to her, turn his torch on and off. The time interval she measures is 15 s. Determine the duration of the torch signal as measured by Mario.

4 A proton inside the Large Hadron Collider travels at 0.999 999 991c. Determine the mass of the proton as measured by a scientist at the LHC and its relativistic momentum.

5 Sera measures her spacecraft and finds that it is 45 m long. She is moving at 0.6c to the right relative to Nazeem; he also measures her spacecraft. Determine the length of the craft as measured by Nazeem.

6 A particle in a laboratory is said to have an approximate lifetime of 3.0×10^{-8} s. When moving at a very high speed in the laboratory, the electron is observed to have a lifetime of 9.8×10^{-8} s. How fast was the electron moving when it was observed to have this lifetime?

7 A proton with mass 1.67×10^{-27} kg moves with speed 0.95c through a potential difference. What is the relativistic momentum of the proton as observed by a lab technician?

11.4 The mass–energy equivalency

WORKED EXAMPLE

Determine the energy required to give an electron a speed of 0.89c if it starts at rest.

ANSWER

$E = (\gamma m - m)c^2$

$E = \left(\sqrt{\frac{1}{1-(0.89c)^2}} \times 9.11 \times 10^{-31}\text{ kg} - 9.11 \times 10^{-31}\text{ kg}\right) \times (3.0 \times 10^8\text{ ms}^{-1})^2$

$E = \left(2.19 \times 9.11 \times 10^{-31}\text{ kg} - 9.11 \times 10^{-31}\text{ kg}\right) \times (3.0 \times 10^8\text{ ms}^{-1})^2$

$E = 1.08 \times 10^{-30} \times (3.0 \times 10^8)^2$

$E = 9.76 \times 10^{-14}\text{ J}$

QUESTIONS

1 If 1 kg of matter could be entirely converted into energy, what would be the value of the energy produced at 14 cents per kWhr?

2 Calculate the rest energy of an electron.

3 Determine the energy required to give an electron a speed of 0.95c if it starts at rest.

4 Can a particle of mass m have total energy less than mc^2? Explain your response.

EVALUATION

Multiple-choice

1 An observer stands between two trees A and B and observes lightning strike the trees simultaneously. The observer standing at the base of tree A would see the lightning strike tree B:

A at the same time the lightning strikes tree A.

B before the lightning strikes tree A.

C after the lightning strikes tree A.

D None of the above

2 As an object with mass moves faster and approaches the speed of light, its relativistic momentum:

A increases.

B decreases.

C remains the same.

D halves.

3 Two spaceships A and B travel at 0.95c towards each other. Spaceship A shines a beacon at spaceship B. Observers on spaceship B observe this beacon to be travelling:

A faster than *c*.

B slower than *c*.

C at *c*.

D None of the above

9780170412643

Short answer

4 Two identical twins are 30-years old when one of them sets out on an extra-terrestrial journey. The twin in the spaceship wears an accurate watch, and when he returns to Earth he claims to be 35-years old. The twin who remained on Earth claims to be 40-years old at this point in time. How fast was the spaceship travelling?

5 Particle A has half the mass of particle B. Particle A also has double the speed of particle B. Does particle A have the same, greater, or less momentum than particle B? Justify your answer.

12 Quantum theory

Summary

- The speed of light in a vacuum is approximately 3.00×10^8 m s^{-1}. Visible light is an electromagnetic wave and the speed of light depends only on the medium.
- The electromagnetic spectrum consists of radio waves, microwaves, infrared light, visible light, ultraviolet light, x-rays and gamma radiation.
- A discrete unit or amount of some physical property, such as energy, charge, mass or angular momentum, may be described as a quantum.
- The oscillating electric and magnetic fields that make up a light wave are coupled.
- When a light wave meets a different medium, the electric and magnetic fields interact with the atoms and electrons in the material, slowing the light wave down and causing its path to refract (bend).
- In Young's double-slit experiment bright fringes (constructive interference) occur along nodal lines where the path difference, δ, is equal to a multiple of the wavelength of light, i.e. $\delta = n\lambda$. Dark fringes (destructive interference) occur along anti-nodal lines where the path difference, δ, differs by exactly one half of the wavelength and the waves are precisely out of phase, i.e. $\delta = \left(n - \frac{1}{2}\right)\lambda$.
- In Young's double-slit experiment, $\Delta y = \frac{n\lambda L}{d}$.
- Constructive interference is the super-positioning of waves where crests intersect with crests and troughs intersect with troughs. Constructive interference is characterised by anti-nodes or bright fringes.
- Destructive interference is the super-positioning of waves where a crest intersects with a trough, due to incoherent wave sources or sources being half a cycle out of phase. This results in a node or dark fringe.
- Wave–particle duality is the dual nature of matter and energy, requiring both wave and particle models to completely explain all observed behaviour of matter and energy.
- The de Broglie wavelength is determined using $\lambda = \frac{h}{p} = \frac{h}{mv}$. The de Broglie wavelength for an electron is given by: $n\lambda = 2\pi r$.
- A continuous spectrum contains radiation of all wavelengths; for example, a rainbow is composed of all the wavelengths of the visible spectrum.
- A black body is an object with a perfectly absorbing surface, which emits black body radiation with a spectrum that is characteristic of the temperature of the object.

9780170412643

- Wien's law determines the position of the peak wavelength of a black body, $\lambda_{max} = \frac{b}{T}$, where λ_{max} = peak wavelength in m, T is the absolute temperature in K, and b is Wein's constant, $b = 2.898 \times 10^{-3}$ m K.
- The energy of a photon(s) of light may be determined using $E = nhf = nh\frac{c}{\lambda}$ where n is an integer, f is the frequency of oscillation in Hz, h is Plank's constant, $h = 6.63 \times 10^{-34}$ J s, $c = 3.00 \times 10^{-8}$ m s^{-1} and λ is the wavelength of light in m.
- The photoelectric effect is the resulting ejection of electrons (photoelectrons) from a surface by incident photons of sufficient energy.
- Experiments with the photoelectric effect apparatus demonstrated that:
  - a photocurrent is only produced when the frequency of the light is above some minimum value, termed the threshold frequency, f_0, regardless of the intensity of the light
  - the number of electrons (the current), if emitted, is proportional to the intensity. It does not vary with the frequency of the light (as long as the frequency is at least the threshold frequency)
  - a unique, minimum amount of energy is required to eject an electron from each metal's surface. This value is known as the work function, $W = hf_0$
  - the kinetic energy of a photoelectron may be determined using $KE_{max} = qV_{stop}$
  - a stopping voltage may be applied to stop a photocurrent – V_{stop} is the reverse bias voltage required to stop the flow of photoelectrons.
- The photoelectric equation: $KE_{max} = hf - hf_0 = E - W$ states that if an electron absorbs light energy hf, and is emitted from the metal, it will have a maximum kinetic energy of $E = hf$, less the energy needed to leave the surface, $W = hf_0$.
- A continuous spectrum is the distribution of components of light or another wave arranged by frequency (or wavelength).
- A line spectrum is an emission or absorption spectrum consisting of discrete lines, characteristic of the energy levels of a particular atom or molecule.
- The absorption spectrum is the wavelengths (or frequencies or energies) of radiation absorbed by a material.
- The emission spectrum is the spectrum of radiation emitted by an object, for example black body radiation or atomic spectra from a discharge tube.
- Rydberg expressed a relationship between electron energy levels and the wavelengths of emitted light through: $\frac{1}{\lambda} = R\left(\frac{1}{n_f^2} - \frac{1}{n_i^2}\right)$, where λ is the wavelength of the line, n_f and n_i are integers and R is a constant known as the Rydberg constant, $R = 1.097 \times 10^7$ m^{-1}.
- Bohr noted several postulates for the Bohr model of the atom.
  - An electron in an atom moves in a circular orbit about the nucleus under the influence of electrostatic attraction.
  - Only certain orbits are stable. Electrons in these orbits do not emit energy.
  - The greater the radius of the orbit, the greater is its energy. Atoms emit radiation when an electron goes from one orbit to another orbit with lower energy. The energy released is found using $E = E_f - E_i = hf$.
  - The orbit of electrons are characterised by quantised radii, given by $r = \frac{nh}{2\pi m_e v}$, where r is the radius in m, m_e is the mass of the electron in kg, v is its velocity in m s^{-1}, h is Planck's constant and n is an integer.
- The angular momentum, L, is the momentum associated with the orbital motion of an electron, $L = mvr$.
- The allowed energies of a nucleus–electron system; often referred to as electron energy levels, are characteristic of the atom.
- For hydrogen, the relationship between wavelength and energy level of an electron is expressed as the Rydberg equation: $\frac{1}{\lambda} = R\left(\frac{1}{n_f^2} - \frac{1}{n_i^2}\right)$, where R = the Rydberg constant, $R = 1.097 \times 10^7$ m^{-1}.

REVISION

12.1 Quantum theory key terms

QUESTIONS

1 Demonstrate your understanding of the nature of light by writing a paragraph that relates all the key terms below.

- Quantum – a discrete unit or amount of some physical property, such as energy, charge, mass or angular momentum.
- Photon – a particle or quanta of light, having energy $E = hf$.
- Wave-particle duality – the dual nature of matter and energy, requiring both wave and particle models to completely explain all observed behaviour of matter and energy.
- Constructive interference – the super-positioning of waves where crests intersect with crests and troughs intersect with troughs.
- Destructive interference – the super-positioning of waves where a crest intersects with a trough, due to incoherent wave sources.
- Spectrum – the distributed components of light or another wave arranged by frequency (or wavelength).

2 Draw a flowchart or mind map that relates and links all the following key terms.

- Black body radiation – the electromagnetic radiation emitted by a black body, with a spectrum characteristic of the temperature of the body.
- Photoelectric effect – the ejection of electrons from a surface by incident photons of sufficient energy.
- Threshold frequency – the minimum frequency of light needed to eject an electron from a metal surface.
- Stopping voltage – the reverse bias voltage required to stop the flow of photoelectrons in a photoelectric experiment.
- Work function – the energy required to eject an electron from a metal surface.

9780170412643

12.2 The nature of light

QUESTIONS

1 The wave model of light has been widely accepted and explains many of the phenomena exhibited by light. There are two experiments in particular that support a different, particle nature of light. Name these two experiments.

2 The speed of electromagnetic radiation in a vacuum was found to be the same as that of light. Based on this, Maxwell suggested that light had two specific properties. State these properties.

3 Construct a table that lists two examples to support both the wave nature of light and the particle nature of light.

4 Match the regions of the electromagnetic spectrum with their corresponding approximate wavelengths.

ELECTROMAGNETIC SPECTRUM REGION	APPROXIMATE WAVELENGTHS (m)
Visible light	$\times 10^{2} - \times 10^{3}$
Radio waves	$\times 10^{9}$
Gamma rays	$\times 10^{-6}$
Ultraviolet light	$\times 10^{0} - \times 10^{1}$
Infrared light	$\times 10^{4} - \times 10^{5}$
X-Rays	$\times 10^{-3}$
Microwaves	$\times 10^{8}$

12.3 Young's double split experiment

QUESTIONS

1 Contrast a nodal point (dark fringe) with an anti-nodal point (bright fringe) in a double-slit experiment.

2 In a double-slit experiment light with wavelength 609 nm is used to illuminate twin slits. The pattern produced is observed on a wall 1.80 m from the slits. The sixth interference minimum (dark spot) was found to be at a position 9.6 mm from the central bright spot. Use this data to calculate the slit separation.

3 In Young's double-slit experiment, describe the effect on the spacing of the light and dark fringes (increases, decreases or stays the same) if:

a the wavelength of light is decreased

b the screen is moved closer to the slits

c the space between the slits is decreased.

4 In a double slit experiment light of wavelength 530 nm is incident on a pair of slits spaced a distance of 1.58 mm apart. If the screen is a distance 2.0 m from the slits, describe the positions of the first three bright spots.

5 In a measurement to determine the wavelength of a light source, a viewing screen is placed a distance 2.4 m from a pair of slits with a separation 0.10 mm. The first dark fringe is a distance of 2.5 cm from the centre line on the screen.

a Determine the wavelength of the light.

b Determine the distance between any two adjacent bright fringes.

12.4 Wave–particle duality of light

QUESTIONS

1 The wave-particle duality of light may be best described as:

A the explanation of the double-slit experiment.

B the dual nature of matter and energy, requiring both wave and particle models to completely explain all observed phenomena.

C the explanation of black body radiation.

D the interference that occurs with light waves.

2 Determine the de Broglie wavelength of a bullet of mass 35 g travelling at $1.0 \times 10^{3}\,\mathrm{m\,s^{-1}}$.

3 A beam of electrons with a de Broglie wavelength of 1.6×10^{-10} m is incident on a double- slit apparatus with a slit separation 3.5×10^{-9} m. If the detectors are arranged on a screen a distance 30 cm from the slits, determine the location of:

a the first interference maximum

b the first interference minimum.

4 A photon of energy 2.8×10^{-15} J collides with a stationary electron that is free to move. After the collision, the photon returns along its original path and the electron moves forwards with a momentum of 1.6×10^{-23} kg m s^{-1}.

a Calculate the magnitude of the momentum of the photon before the collision.

b What is the de Broglie wavelength of the electron after the collision?

c Calculate the wavelength of the photon before and after the collision.

12.5 Black-body radiation

QUESTIONS

1 Define the following terms.

a Absorption spectrum

b Continuous spectrum

c Emission spectrum

2 Describe the characteristics of a black body.

3 Describe the relationship between the peak wavelength emitted from a black-body and its surface temperature. Use this relationship to determine the surface temperature of peak wavelength 4.05×10^{-7} m (405 nm). Use: $b = 2.898 \times 10^{-3}$ m K.

4 An incandescent light globe is connected to a variable power supply and the voltage gradually increased. Describe the sequence of colours produced by the filament of the globe. What is the peak wavelength emitted by a toaster element at 700°C? What colour would you expect it to be?

12.6 Planck's quanta and photon characteristics

QUESTIONS

1 Demonstrate that the units for Planck's constant must be J s using the formula for energy.

2 An atomic oscillator has a frequency $f = 5.8 \times 10^{12}$ Hz, and is in the $n = 3$ state.

a What is the energy of this oscillator?

b What frequency of light will be emitted if it transitions to the $n = 2$ state?

______

______

3 Define the following terms.

a Emission spectrum

______

______

b Energy level

______

______

c Quanta

______

______

4 The energy of an electron is quantised. Describe in your own words what this means.

______

______

______

______

12.7 The photoelectric effect

QUESTIONS

1 Describe how a photoelectron differs from any other electron.

______

______

2 It requires more energy to remove an electron from the surface of a polished piece of silver than from a polished piece of copper.

a Which metal has the larger work function?

______

______

b Which metal has the greater threshold frequency?

____

____

3 The graph in Figure 12.7.1 shows the results of a photoelectric experiment using magnesium metal.

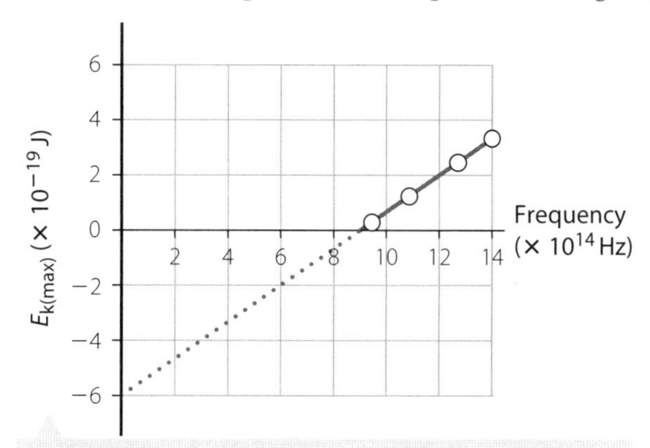

FIGURE 12.7.1 A plot of photoelectron maximum kinetic energy as a function of frequency of incident light for magnesium

a Determine a value for Planck's constant from this graph.

____

b Determine a value for the work function of magnesium.

____

c Imagine that silver had been used in this experiment instead of magnesium. Silver has a work function of 4.70 eV. Where on the graph would a line representing silver exist?

____

4 A polished lead surface is illuminated with light of wavelength 190 nm, prompting the ejection of electrons. What is the effect on the photocurrent if:

a the wavelength is halved?

____

____

b the intensity of the light is doubled?

____

____

5 Figure 12.7.2 shows a photoelectric tube with light of frequency f and intensity I incident on a metal cathode. Electrons emitted from the cathode are collected at the anode. The potential difference between the anode and cathode is varied, and the resulting photocurrent is measured. Figure 12.7.3 shows the results of this experiment.

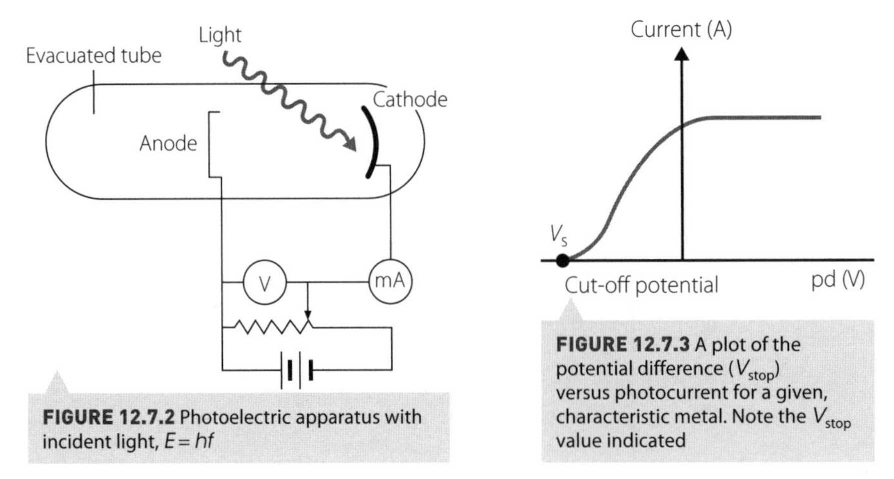

FIGURE 12.7.2 Photoelectric apparatus with incident light, $E = hf$

FIGURE 12.7.3 A plot of the potential difference (V_{stop}) versus photocurrent for a given, characteristic metal. Note the V_{stop} value indicated

a Why is the photocurrent constant at positive values of potential difference?

---

b If the frequency of the light is varied, which of the graphs in Figure 12.7.4 represent the relationship between the stopping voltage, V_s, and f?

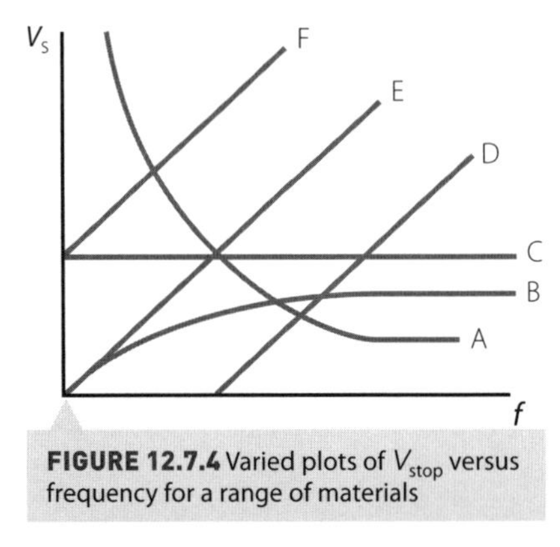

FIGURE 12.7.4 Varied plots of V_{stop} versus frequency for a range of materials

---

9780170412643

6 Figure 12.7.5 shows a graph of maximum kinetic energy as a function of frequency for a particular metal. Copy the graph and, on the same set of axes, draw a line showing the maximum kinetic energy for light of the same intensity but for a metal with a greater work function.

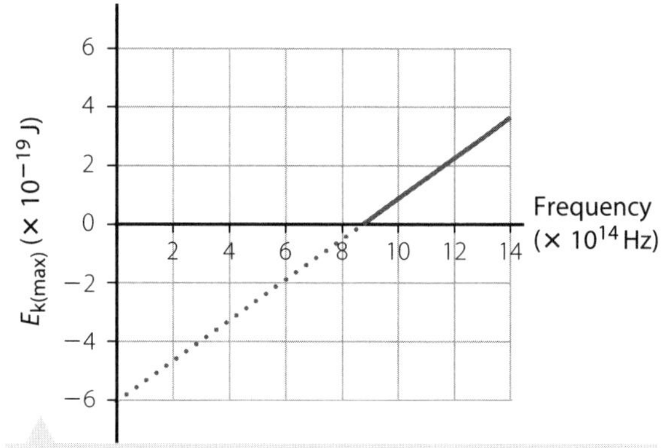

FIGURE 12.7.5 A plot of kinetic energy versus frequency for photoelectrons

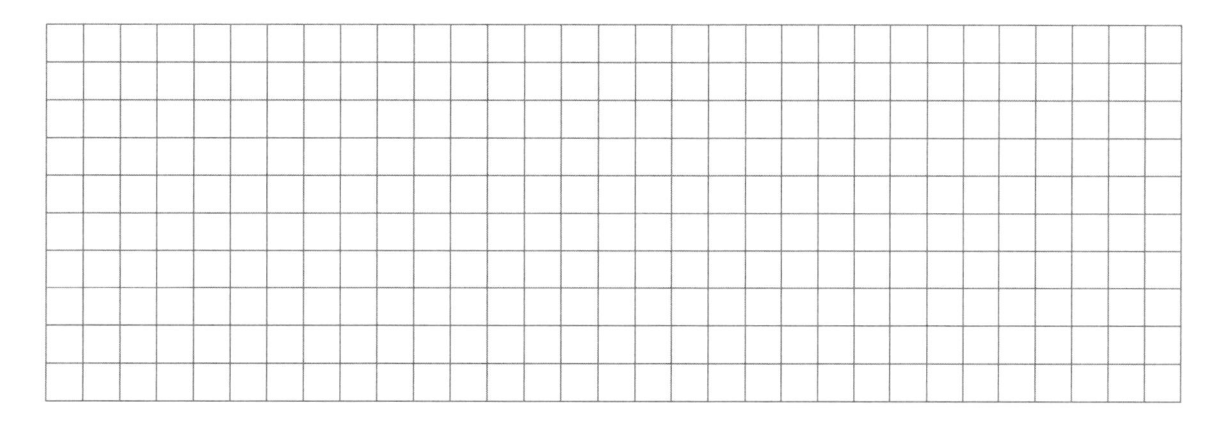

7 A polished sodium surface with a work function of 2.28 eV is illuminated with light.

a Determine the threshold wavelength for sodium. What colour does this correspond to?

____

____

b Determine the maximum kinetic energy of ejected photoelectrons when light of wavelength 300 nm is used.

____

____

8 The table below shows data collected in a photoelectric experiment.

WAVELENGTH (nm)	FREQUENCY ($\times 10^{14}$ HZ)	KE_{max} OF PHOTOELECTRONS (eV)	KE_{max} OF PHOTOELECTRONS (J)
647		0.74	
556		1.08	
490		1.49	
439		1.79	

Complete the table and plot an appropriate graph to determine:

a Planck's constant

b the work function for this metal.

12.8 The model of the atom and atomic spectra

QUESTIONS

1 State Bohr's four postulates that are the basis of his atomic model.

2 The spectral lines for transitions to the $n = 2$ state from higher levels for a particular atom are found to be in the visible light region of the electromagnetic spectrum. Would this atom be expected to have any spectral lines in the ultra-violet range? If so, to what transitions would they correspond?

3 Calculate the wavelength and frequency of an X-ray photon of energy 2.8×10^4 eV.

4 Two energy levels within a particular atom are 8.60 eV and an unknown x eV. When an electron of this element returns from the higher level to the lower energy level, radiation of wavelength 550 nm is emitted. Determine the value of x.

5 An atom is excited to its fourth energy level above the ground state, that is $n = 5$.

a How many different spectral lines can it emit?

b Which of the energy level transitions will produce the photon of greatest energy?

c Which of the energy level transitions will produce a photon of the longest wavelength?

Multiple-choice

1 The electromagnetic spectrum consists of several regions. Which of the options below lists the electromagnetic regions in order of increasing wavelength?

A Ultraviolet, infrared, gamma rays and X-rays

B Infrared, gamma rays, X-rays and ultraviolet

C Infrared, ultraviolet, gamma rays and X-rays

D Gamma rays, X-rays, ultraviolet and infrared

2 What series of the hydrogen emission spectra emits light in the visible region?

A Lyman series

B Paschen series

C Balmer series

D Rydberg series

3 Which statement regarding two wave interference is *incorrect*?

A Constructive interference occurs along anti-nodal lines.

B Destructive interference occurs when a wave crest meets with a wave trough.

C A wave crest meeting a wave trough is an example of constructive interference.

D Nodal lines always result in a dark fringe.

Short answer

4 A black body is known to have a surface temperature of 3800 K. Use the value of $b = 2.898 \times 10^{-3}$ m K to determine the peak wavelength.

____

____

____

5 The filament of an incandescent light globe can be modelled as a black body. A tungsten filament reaches a temperature of 3100 K. Determine its peak wavelength.

6 In a double-slit experiment, light with wavelength 589 nm is used to illuminate twin slits that are separated by 0.015 mm. The pattern produced is observed on a wall 1.20 m from the slits. Determine the position of:

a the first interference maximum (bright spot)

b the first interference minimum (dark spot).

7 Describe the features of a black body.

8 A pair of slits spaced 0.015 mm apart are illuminated with light of two wavelengths at the same time: $\lambda_1 = 630$ nm and $\lambda_2 = 420$ nm. The viewing screen is a distance 2.0 m from the slits. At what position on the screen, other than at $y = 0$, do the maxima from the two interference patterns first coincide?

9 Predict the outcome of shining light of frequency $f = 4.25 \times 10^{15}$ Hz onto an iron metal surface of work function 7.25×10^{-19} J.

10 Determine the de Broglie wavelength of an electron travelling at $18.0\,m\,s^{-1}$. State what occurs to the wavelength of an electron as the velocity is increased.

11 A quantum of energy has a wavelength 4.25×10^{-5} m.

a Determine the frequency of this quantum.

b Determine the energy of this quantum.

c State its energy in eV.

13 The Standard Model

Summary

- Elementary particles are those that have no internal structure and cannot be divided into smaller components.
- The electron is one of a family of elementary particles called leptons.
- The photon is an elementary particle with zero mass.
- Protons and neutrons are not elementary particles. They can decay into other particles. Neutrons have a magnetic field even though they are uncharged.
- For every particle there is a corresponding antiparticle with opposite charge but the same mass. When a particle meets its antiparticle they annihilate, producing energy in the form of photons.
- High-energy particles can be detected because they cause ionisation of materials they pass through. This occurs in cloud and bubble chambers.
- Particle accelerators are used to create new particles by colliding high-energy charged particles with a target, such as another particle or a nucleus.
- Particles are grouped according to their intrinsic properties including: mass, lifetime, spin, baryon number and lepton number.
- Spin is a quantum property of a particle resulting from its magnetic moment.
- The Pauli exclusion principle states that particles with half-integer spin cannot occupy identical sets of quantum numbers.
- Fermions are particles with half-integer spin, while bosons are particles with integer spin.
- Leptons are fundamental particles with no internal structure, while hadrons are large mass particles containing a composite internal structure made up of quarks.
- Gauge bosons are force carrying particles that mediate particle interactions through the four fundamental forces.
- The four fundamental forces are the strong force (not the same as the strong nuclear force), the weak force, the electromagnetic force and the gravitational force.
- The gauge bosons are: photons for the electromagnetic force, W and Z bosons for the weak force and gluons for the strong force, the gravitational force has been hypothesised to be due to an exchange of gravitons, although these are yet to be observed.
- The strong nuclear force, which is mediated by pions, is a result of the strong force interaction between quarks, which is mediated by gluons.
- Leptons are fundamental particles (for example the electron) that have a low mass and no internal structure They do not interact via the strong nuclear force or the strong force.

- There are six leptons: the electron, the electron neutrino, the muon, the muon neutrino, the tau and the tau neutrino.
- Hadrons are a family of large mass particles that have an internal structure made up of quarks and interact via all fundamental forces.
- Quarks are a family of fundamental particles that are never observed in isolation and join together to produce mesons and baryons. They contain fractional charges of $\frac{1}{3}$ or $\frac{2}{3}$.
- Mesons are composed of a quark-antiquark pair. Baryons are composed of three quarks.
- There are six 'flavours' of quark: up, down, strange, charm, top and bottom. Quarks contain fractional charges of $\frac{1}{3}$ or $\frac{2}{3}$.
- Limitations of the standard model include:
  - the lack of a description for gravity
  - the prediction that neutrinos should be massless
  - the prediction of magnetic monopoles.
- It has been instrumental in the development of the Big Bang theory and the early universe. In the first stages of the universe, the four forces were unified as a single force.

9780170412643

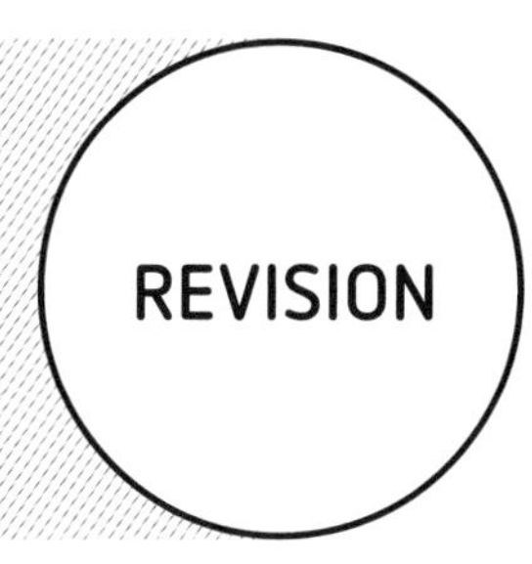

13.1 Elementary particles and antiparticles

Experimental evidence has been provided by particle accelerators and from cosmic rays that indicates the existence of elementary particles, particles that have no internal structure and cannot be divided into smaller components. Under the standard model of particle physics, there are three broad categories of elementary particles; quarks, leptons and gauge bosons. In addition, each elementary particle has an antiparticle that has the same mass but opposite charge. When a particle-antiparticle pair come into contact, they undergo annihilation, producing energy in the form of photons.

QUESTIONS

1 Complete the following statements by filling in the spaces provided.

a The ancient ________________ believed that the universe was made out of four fundamental ________________: earth, air, fire and water.

b ________________ hypothesised that matter was composed of indivisible ________________, which were later shown to be composed of even more fundamental components including ________________, ________________ and ________________.

c The ________________ model of particle physics uses experimental evidence from particle ________________ and cosmic rays to predict that these components were themselves made from smaller components termed ________________ particles.

d An elementary particle is one which has no internal ________________ and cannot be ________________ into smaller components.

2 At the top of the figure below, the paths of an electron and a positron, spiralling inwards, are seen. A uniform magnetic field is directed into the page.

a Which is the electron and which the positron? Explain your reasoning.

b Both particles started with the same velocity. As they spiralled inwards, which slowed more rapidly?

c What is the equation to calculate the magnitude of the force required to move a body of mass, m, at speed, v, in a circular path of radius, r?

d What is the equation to calculate the magnitude of the force acting on a body with charge, q, moving at speed, v, in a region of uniform magnetic field, B.

e Use your answers to parts **c** and **d** to determine an expression for the momentum of that body.

9780170412643

13.2 Particle physics: the continuing search for elementary particles

So far, particle accelerator experiments have discovered hundreds of particles that can be grouped into a number of classifications. Leptons are a family of fundamental particles. Baryons are a family of heavy subatomic particles that contain composite structures made up of quarks. Some properties that can be used to differentiate these particles include: mass, length of lifetime, spin, lepton number or baryon number. The spin of a particle is a quantum property resulting from its magnetic moment. Fermions are particles that have half-integer spin and bosons are particles with integer spin. Fermions obey the Pauli exclusion principle meaning that no two fermions can have identical sets of quantum numbers.

QUESTIONS

1 Decide whether each of the following descriptions is indicative of fermions, bosons, leptons, hadrons, baryons or mesons.

a The groups of particles that have the smallest mass.

____

b The group of particles that have zero spin.

____

c The group of particles that have spin $\frac{1}{2}$.

____

d The group of particles that have integer spin.

____

e The group of particles that tend to be the most stable.

____

f The particles that have the shortest lifetime.

____

g The particles that have baryon number zero.

____

2 Determine the identity of the particle which has the following properties.

a A stable lepton with a L_e number of +1 and a mass of 0.511 MeV c^{-2}.

____

b The heaviest and shortest lived lepton.

____

c A meson with a mass of 493.7 MeV c^{-2}.

____

d A meson which is its own anti-particle.

____

e A baryon with a spin of $\frac{3}{2}$ and a mass of 1672 MeV c^{-2}.

____

f A stable hadron with a baryon number of +1.

____

13.3 Gauge bosons and the fundamental forces of nature

The third family of particles are the gauge bosons. These are force-carrying particles that mediate particle interactions through the four fundamental forces: the electromotive force, the weak force, the strong force and gravity. In the particle exchange model, when two particles interact via one of the four fundamental forces, a gauge boson is emitted by one particle and then absorbed by the other. The gauge bosons are: photons for the electromagnetic force, W and Z bosons for the weak force and gluons for the strong force, the gravitational force has been hypothesised to be due to an exchange of gravitons, although these are yet to be observed. The strong nuclear force, which is mediated by pions, is a result of the strong force interaction between quarks, which is mediated by gluons.

QUESTIONS

1 Complete the following statements regarding the fundamental forces of nature and gauge bosons.

a A ______ boson is a ______ carrying particle that mediates particle ______ through the four fundamental forces.

b In the ______ exchange model of interactions, one particle ______ a gauge boson that is subsequently ______ by another particle.

c The electromagnetic force is said to mediated by the ______.

d The ______ ______ force operates as a force of ______ between nucleons.

e The ______ force operates as a force of attraction between ______.

f The weak nuclear force is involved in ______ decay and is mediated by ______ and ______ bosons.

g The ______ theory postulates that the ______ and ______ interactions are unified at high ______.

h The ______ is the gauge boson which has been hypothesised to mediate the ______ force.

2 Complete the following table for the properties of gauge bosons.

INTERACTION	RELATIVE STRENGTH	RANGE	GAUGE BOSON	MASS (GeV c^{-2})
Strong				
Electromagnetic				
Weak				
Gravitational				

9780170412643

13.4 Leptons

Leptons are fundamental particles that have a low mass and no internal structure. They do not interact via the strong nuclear force or the strong force. There are six leptons: the electron, the electron neutrino, the muon, the muon neutrino, the tau and the tau neutrino.

QUESTION

1 Complete the following table for the properties of leptons.

NAME	SYMBOL	ANTI-PARTICLE	MASS ($MeV c^{-2}$)	B	L_o	L_u	L_T	LIFETIME (s)	SPIN
Electron									
Electron-neutrino									
Muon									
Muon-neutrino									
Tau									
Tau-neutrino									

13.5 Hadrons: mesons, baryons and their quarks

Hadrons are a family of large mass particles that have an internal structure made up of quarks and interact via all of the fundamental forces. Quarks are a family of fundamental particles that are never observed in isolation and join together to produce mesons and baryons. Mesons are composed of a quark-antiquark pair. Baryons are composed of three quarks. There are six 'flavours' of quark: up, down, strange, charm, top and bottom. Quarks contain fractional charges of $\frac{1}{3}$ or $\frac{2}{3}$.

QUESTIONS

1 Complete the following table for the properties of quarks.

NAME	SYMBOL	SPIN	CHARGE	BARYON NUMBER	STRANGENESS	CHARM	BOTTOMNESS	TOPNESS
Up								
Down								
Strange								
Charmed								
Bottom								
Top								

2 Fill in the following diagrams to indicate the quarks that are required to compose the given particles.

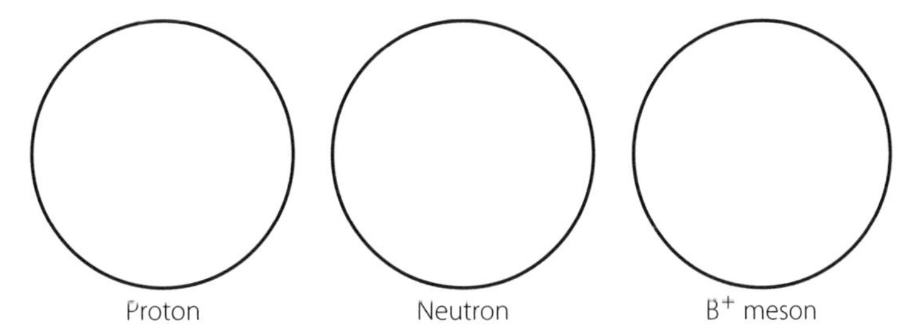

13.6 The Standard Model today

The standard model has correctly predicted the existence of many new particles including the Higgs boson, but it does have its limitations. It does not include a description for gravity and it predicts that a neutron should be massless. It has been instrumental in the development of the Big Bang theory and the early universe. In the first stages of the universe, the four forces were unified as a single force.

QUESTIONS

1 Place the named particles in the Venn diagram to indicate which of the fundamental force they interact with or mediate:

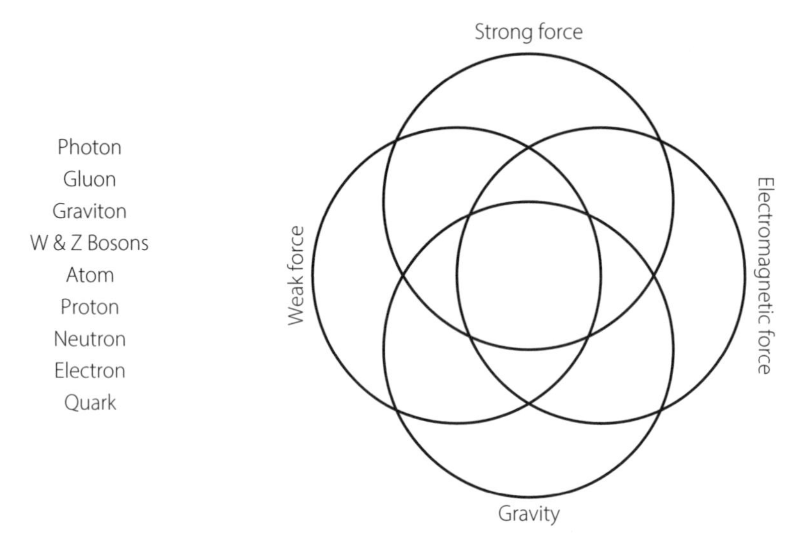

2 Complete the following history of the universe diagram.

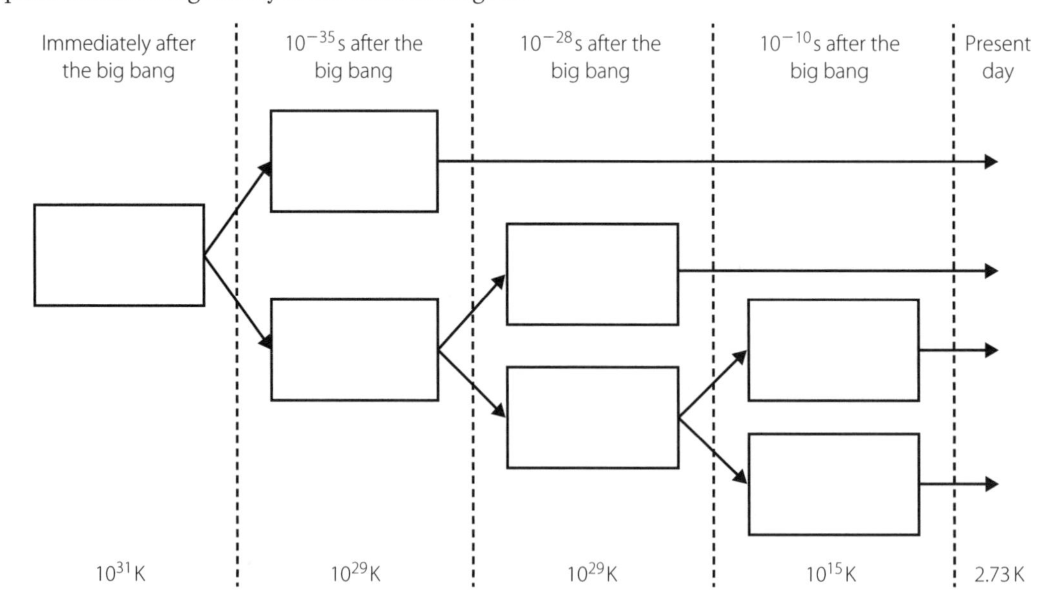

9780170412643

3

Across:

1 This property must be conserved in the Standard Model

7 Composed of uud quarks

8 Should be massless but isn't

10 A meson or baryon

11 Most of the universe is made of this

14 An intriguing quark flavour

17 We live in the milky one

18 This group includes positrons

20 Early on, an energetic photon would hit a newly-formed H atom and this would happen

21 Proportions of matter and antimatter are not

Down:

1 A type of field particle

2 Receding galaxies show this shift in their wavelengths

3 A cheerful quark flavour

4 We now know that it's increasing

5 Can carry away any excess energy after a reaction

6 Delta particles have $\frac{3}{2}$

9 The whole universe is doing this

10 Particles interact with this to acquire mass

12 Can be converted into mass

13 A delightful quark flavour

15 I stick quarks together

16 Occur in twos or threes; not alone

19 A type of neutrino

Multiple-choice

1 Which of the following would be considered an elementary particle?

A An alpha particle

B A beta particle

C A neutron

D All of the above

2 Which of the following would *not* have half-integer spin?

A Fermions

B Leptons

C Baryons

D Mesons

3 Which of the following is *not* a lepton?

A An electron

B A positron

C A pi meson

D A muon

4 Which of the following is correct?

A The strong force is mediated by pion exchange.

B The strong nuclear force is mediated by pion exchange.

C Both the strong force and the strong nuclear force are mediated by pion exchange.

D The weak force is mediated by pion exchange.

5 Which of the following is considered evidence of the Big Bang theory?

A Immense clouds of hydrogen

B Cosmic rays

C Microwave background radiation

D The universe acting as a black body

6 What is dark matter?

A Objects that absorb all or most of the radiation they receive.

B Non-luminous objects, such as planets and moons.

C Hypothesised mass-energy that repels normal matter, increasing the expansion of the universe.

D The missing matter in the universe.

Short answer

7 Name the three families of fundamental particles described in this chapter.

---

8 State the fundamental force that is yet to be fully described by the standard model of particle physics and the particle which has been hypothesised to be mediated by it?

---

9 Compare the relative strengths of the electromagnetic force and the gravitational force.

---

10 Explain why the standard model is considered incomplete.

9780170412643

11 Deduce the identity, spin, charge and baryon number of a baryon that consists of two up quarks and a down quark.

12 a Calculate the electrostatic and the gravitational attraction between a proton and an electron which are 5.3×10^{-11} km apart (the most probable distance apart in a hydrogen atom).

b Which force is stronger and how many times stronger is it?

13 What is the strongest force acting in a deuterium (Hydrogen-2) atom?

14 Particle interactions

Summary

- Leptons and baryons are both fermions which obey the Pauli exclusion principle.
- Lepton numbers are the quantum numbers associated with each lepton, antilepton and non-leptonic particles. L = +1 for leptons, L = −1 for antileptons and L = 0 for non-leptonic particles.
- Baryon numbers are quantum number associated with each baryon, antibaryon and non-baryonic particles. B = +1 for baryons, B = −1 for antibaryons and B = 0 for non-baryonic particles.
- Lepton and baryon numbers are always conserved in particle reactions. If these numbers are not conserved, then it is not an allowed reaction.
- Reaction diagrams are a pictorial representation of a particle interaction. Arrows represent particles, and the direction of the arrows match the direction of time where time is on the horizontal axes. Antiparticles point in the opposite direction of time.
- Exchange particles are particles carrying force which are responsible for behaviour during particle interactions. They are sometimes formed as a result of a particle interaction.
- Feynman diagrams are used to show detailed particle interactions with various particles and exchange forces. The following rules apply to Feynman diagrams.
  - Time is measured in the positive y direction.
  - Space is measured in the positive x direction.
  - Solid straight lines with upward arrows represent particles.
  - Solid straight lines with downward arrows represent antiparticles.
  - Wavy lines represent electromagnetic force from exchange particles (photon).
  - Helical lines represent strong force from exchange particles (gluon).
  - Dashed lines represent weak force from exchange particles (boson).
  - Charge is always conserved during the interaction.
  - Baryon number is always conserved during the interaction.

9780170412643

- There are three types of symmetries in particle interactions which model physical phenomena under transformations of space or time. These symmetries are used to predict new reactions.
  - Time-reversal symmetry is when reactions are reversed in time and still obey all the conservation laws. This means that the reaction diagram is reflected, swapping left to right.
  - Charge-reversal symmetry says that if the charges on all particles in a reaction are reversed, then this new reaction is also possible in that it does not violate any conservation laws. Charge reversal switches all particles to antiparticles and vice versa.
  - Crossing symmetry is when one particle is taken to the opposite side of the reaction and converted to its antiparticle while not violating any conservation principles.
- Symmetry breaking is a change in the behaviour of a physical system or the laws of physics that govern its behaviour when a symmetry operation such as a translation, reflection or rotation in time or space takes place.
- While symmetries may be theoretically possible, sometimes the outcome is almost never observed experimentally.

REVISION

14.1 Conservation of lepton and baryon number

QUESTIONS

1 What is the difference between a baryon and a lepton?

____

____

2 State whether the following reactions possible under conservation of lepton and baryon number:

a $\pi^- + p \rightarrow K^- + \Sigma^0$

____

b $\mu^- \rightarrow e^- + \nu_e + \nu_\mu$

____

c $n \rightarrow p + e^- + \nu_e$

____

d $\Omega^- \rightarrow \Lambda^0 + \Sigma^-$

____

e $n \rightarrow p + e^- + \bar{\nu}_e$.

____

14.2 Reaction diagrams and Feynman diagrams

WORKED EXAMPLES

1 An antimuon decays into a positron, electron neutrino and muon-antineutronio. Draw the reaction diagram for this process.

2 A neutron decays into a proton an electron during $\beta-$ decay. The proton and electron interact weakly after this process, and an electron antineutrino is also formed in this process. Represent this interaction with a Feynman diagram.

ANSWERS

1

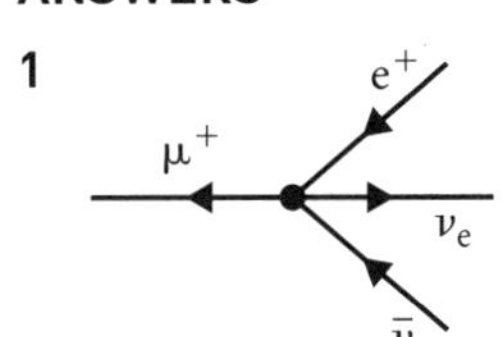

2

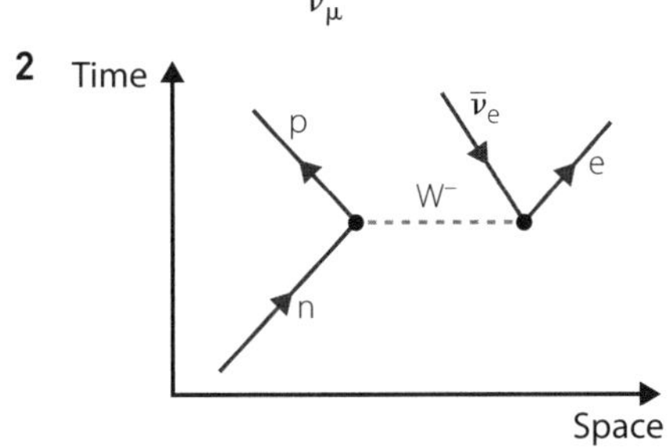

QUESTIONS

1 Show the decay of a neutron with a reaction diagram.

2 Represent the reaction $\Sigma^- \rightarrow n + \pi^-$ with a reaction diagram.

3 Represent a proton capturing an electron to form a neutron and electron neutrino as a reaction diagram.

4 A proton and an electron interact weakly to produce a neutron and an electron neutrino. Represent this interaction with a Feynman diagram.

5 Represent electron-positron annihilation with a Feynman diagram if the type of interaction they have is a particle interaction to produce two high energy photons

6 A muon decays into a muon neutrino, electron and electron antineutrino. The muon neutrino interacts weakly with the electron and electron antineutrino pair. Represent this reaction with a Feynman diagram.

9780170412643

14.3 Symmetry

WORKED EXAMPLES

1 Apply time reversal symmetry to $\beta-$ decay and state whether the reaction is likely to be observed. Represent this time reversal with a reaction diagram.

2 Apply charge reversal symmetry to the decay of a muon $\mu^- \rightarrow e^- + \bar{\nu}_e + \nu_\mu$ and write the new equation.

3 Apply crossing symmetry to Σ^- decay if Σ^- commonly decays by producing a neutron and a negative pion. Represent this crossing symmetry with a reaction diagram to predict how a neutron can be formed.

ANSWERS

1 β^- decay:

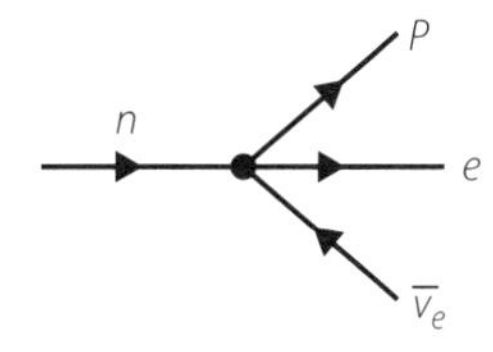

Time reversal:

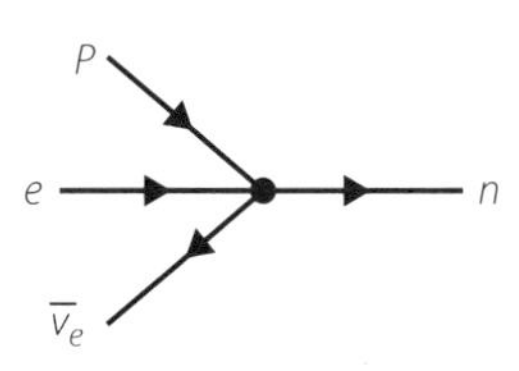

2 $\mu^- \rightarrow e^- + \bar{\nu}_e + \nu_\mu$

Converting charges and changing particles to their antiparticles we obtain:

$\mu^+ \rightarrow e^+ + \nu_e + \bar{\nu}_\mu$

3

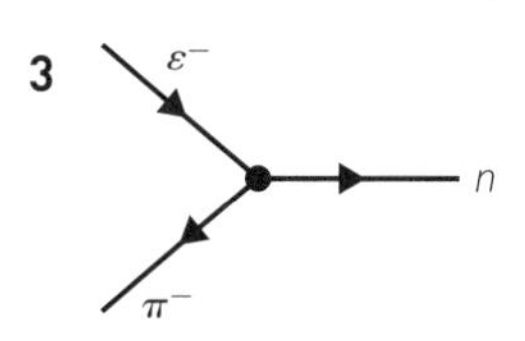

QUESTIONS

1 Apply time reversal symmetry to the reaction shown below and draw the new reaction diagram.

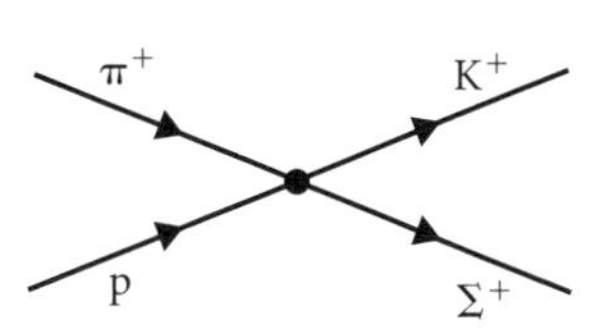

2 Crossing symmetry may be applied to the reaction shown in Question 1 above. To represent a reaction that might actually occur, which of the particles should we choose to move from the right to the left hand side of the reaction? Explain your answer and draw the new reaction diagram.

3 Apply charge reversal symmetry to the reaction of antimuon decay shown in Question 1 above and draw the new reaction diagram.

4 Apply time reversal symmetry to $\Sigma^- \rightarrow n + \pi^-$ and write the new equation.

5 Apply charge reversal symmetry to kaon decay, if a kaon particle decays into two positive pions and a negative pion. Do you think this is likely to occur?

6 Apply crossing symmetry to the reaction $\tau^- \rightarrow e^- + \nu_\tau + \bar{\nu}_e$ to show how an electron and electron antineutrino can possibly form. Write the new equation.

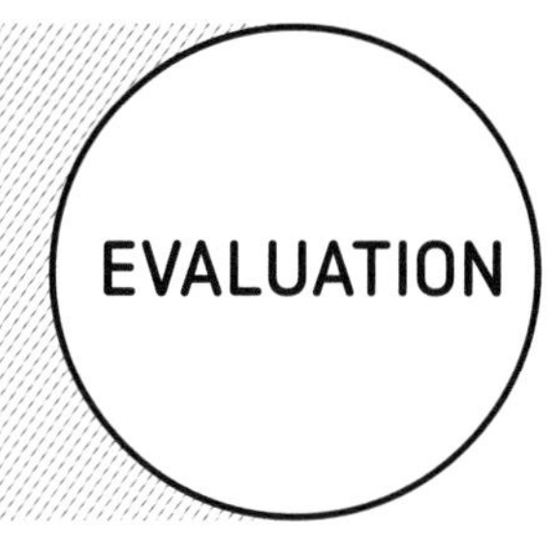

Multiple-choice

1 In particle interactions, what quantities must always be conserved?

A Baryon number and lepton number

B Baryon number, lepton number, spin and electric charge

C Baryon number, lepton number, spin, electric charge, linear momentum and angular momentum

D Baryon number, lepton number, spin, electric charge, linear momentum, angular momentum and mass

2 In a Feynman diagram, what does the helical line represent?

A Electromagnetic force from electrons

B Strong force from gluons

C Weak force from bosons

D Gravitational force from gravitons

3 If charge reversal is applied to an electron antineutrino, the particle obtained is:

A an electron neutrino.

B a positron neutrino.

C a positron antineutrino.

D an electron antineutrino.

Short answer

4 Is the reaction $\mu^- \rightarrow e^- + \bar{\upsilon}_e + \upsilon_\mu$ allowed? Explain.

---

5 Draw the Feynman diagram for β^+ decay.

6 After applying symmetry to a reaction, is the new reaction always observable? Explain.

9780170412643

PRACTICE EXAM BOOKLET

PHYSICS UNITS 3 & 4

MULTIPLE-CHOICE QUESTIONS

Question 1

For a projectile that is launched at an angle of elevation of 25° and velocity of 14.0 m s^{-1}, the horizontal component and vertical component of the launch velocity are, respectively:

A 5.92 m s^{-1}, 12.7 m s^{-1}.

B 24.25 m s^{-1}, 5.92 m s^{-1}.

C 12.7 m s^{-1}, 24.25 m s^{-1}.

D 12.7 m s^{-1}, 5.92 m s^{-1}.

Question 2

The resolutes of the force of weight parallel and force of weight perpendicular on an object placed on an inclined plane are represented, respectively, by:

A $mg\cos\theta$, mg.

B $mg\sin\theta$, $mg\cos\theta$.

C $mg\sin\theta$, $mg\cos\theta$.

D mg, $mg\sin\theta$.

Question 3

The formula for force centripetal and force gravitational may be used to derive an equation for the orbit velocity of a satellite. This equation for determining the orbit velocity of a satellite is:

A $v = gr$.

B $v = mgr$.

C $v = \sqrt{gr}$.

D $v = \frac{mg}{r^2}$.

Question 4

Which of the following equations exhibit an inverse square relationship?

A $F = ma$

B $F = \frac{kqQ}{x^2}$

C $y = \frac{1}{x}$

D $A = \pi r^2$

Question 5

A ball is thrown upwards from the top of a 50 m high cliff at a speed of $8.0\,m\,s^{-1}$ and angle of 15°. The distance from the bottom of the cliff that the ball will land is:

A 7.73 m.

B 3.41 m.

C 61.82 m.

D 26.35 m.

Question 6

A 450 g mass is stationary on a plane while the plane is gently lifted up. At an incline of 15° the mass begins to slide down with an acceleration of $0.84\,m\,s^{-2}$. The static and kinetic friction values, respectively are:

A 1.14 N, 0.76 N.

B 0.76 N, 1.14 N.

C 1.14 N, 1.14 N.

D 0.84 N, 0.76 N.

Question 7

The central tenet of the scientific method whereby hypotheses are tested using observation and experimentation is also referred to as:

A conservation of energy.

B an authoritative method.

C an empirical method.

D fair testing.

Question 8

A 400 g mass placed on a 40° frictionless surface experiences an acceleration of:

A 2.52 N.

B $1.01\,m\,s^{-2}$.

C $6.30\,m\,s^{-2}$.

D $2.52\,m\,s^{-2}$.

Question 9

The gravitational field model explains why objects may exert forces at a distance. The model also allows us to:

A predict the acceleration of an object within a gravitational field.

B calculate the mass of an object from the observed force that it exerts on another object.

C calculate the mass of distant objects, such as planets, by observing their orbits about the Sun.

D All of the above

Question 10

In SI units, the period, frequency, mass and velocity respectively are measured in the units:

A s, kg, $km\,h^{-1}$, Hz

B min, kg, $km\,h^{-1}$, Hz

C s, Hz, kg, $m\,s^{-1}$

D s, kg, $m\,s^{-1}$, Hz

9780170412643

Question 11

In a double slit experiment light of wavelength 530 nm is incident on a pair of slits spaced a distance of 1.58 mm apart. If the screen is a distance 2.0 m from the slits, the second bright spot is located at:

A 6.71×10^{-4} m.

B 1.34×10^{-3} m.

C 2.68×10^{-3} m.

D 3.35×10^{-4} m.

Question 12

How far away from a wire carrying a current of 20 A will the B field have a magnitude of 5.0×10^{-4} T?

A 5.09×10^{-3} m

B 0.01 m

C 8.00×10^{-3} m

D 2.55×10^{-3} m

Question 13

The regions of the electromagnetic spectrum listed in order of increasing wavelength are best represented as:

A infrared light → gamma rays → ultraviolet light → X-rays, microwaves → visible light → radio waves.

B gamma rays, X-rays → ultraviolet light → visible light, infrared light, microwaves → radio waves.

C visible light → radio waves → infrared light → X-rays → gamma rays → ultraviolet light → microwaves.

D visible light → radio waves → gamma rays → ultraviolet light → infrared light → X-rays → microwaves.

Question 14

Two charges $q = 11\,\mu C$ and $Q = 8\,\mu C$ are separated by a distance of 0.08 m. Determine the magnitude of force exerted by q on Q.

A 123.8 N

B 1.375×10^{-8} N

C 9.90 N

D 0.9 N

Question 15

Determine the de Broglie wavelength of a bullet of mass 35 g travelling at 1.0×10^{3} m s^{-1}.

A 2.32×10^{-38} m

B 6.63×10^{-37} m

C 1.89×10^{-35} m

D 1.82×10^{-6} m

Question 16

An electromagnetic wave of frequency $f = 1.65 \times 10^{18}$ Hz is travelling in a vacuum. The wavelength of the electromagnetic wave is:

A 5.50×10^{9} m.

B 5.50 nm.

C 1.82×10^{-6} m.

D 18.2 nm.

Question 17

Which statement regarding two wave interference is *incorrect*?

A Nodal lines always result in a dark fringe.

B Destructive interference occurs when a wave crest meets with a wave trough.

C A wave crest meeting a wave trough is an example of constructive interference.

D Constructive interference occurs along anti-nodal lines.

Question 18

An electron is ejected into a *B* field with velocity $2.0 \times 10^{5}\,\mathrm{m\,s^{-1}}$. It experiences a force of $3 \times 10^{-2}\,\mathrm{N}$ downwards when entering the field. Determine the magnitude of the *B* field.

A $1.07 \times 10^{-12}\,\mathrm{T}$

B $9.38 \times 10^{11}\,\mathrm{T}$

C $6.67 \times 10^{6}\,\mathrm{T}$

D $33.3\,\mathrm{T}$

Question 19

The weak nuclear force is involved in *X* decay and is mediated by *Y*. *X* and *Y* in this sentence represent:

A X = beta decay, Y = force carrier bosons.

B X = alpha decay, Y = fermions.

C X = beta decay, Y = gamma rays.

D X = lepton decay, Y = leptons.

Question 20

A polished sodium surface with a work function of 2.28 eV is illuminated with light. The threshold wavelength for sodium and the maximum kinetic energy of ejected photoelectrons when light of wavelength 300 nm is used are, respectively:

A $1.83 \times 10^{-7}\,\mathrm{m}$, $1.49 \times 10^{-19}\,\mathrm{J}$.

B $5.45 \times 10^{-7}\,\mathrm{m}$, $2.98 \times 10^{-19}\,\mathrm{J}$.

C $1.09 \times 10^{-6}\,\mathrm{m}$, $1.03 \times 10^{-18}\,\mathrm{J}$.

D $109\,\mathrm{nm}$, $1.39 \times 10^{-18}\,\mathrm{J}$.

SHORT-RESPONSE QUESTIONS

Question 1

Perform calculations to complete the table below for an object launched across level ground, neglecting air resistance.

LAUNCH VELOCITY ($\mathrm{ms^{-1}}$)	LAUNCH ANGLE (DEGREES)	HORIZONTAL COMPONENT OF INITIAL VELOCITY ($\mathrm{m\,s^{-1}}$)	VERTICAL COMPONENT OF INITIAL VELOCITY ($\mathrm{m\,s^{-1}}$)	HORIZONTAL RANGE (m)
20	30			
30	45			
40	25			
50	60			

Question 2

To what velocity must a satellite be propelled if it is to maintain an orbit with a radius of 500 km around Earth? Use:

- $M_E = 5.97 \times 10^{24}$ kg
- $r_E = 6.37 \times 10^6$ m

Question 3

Find the east-west and the north-south resolutes of a vector of magnitude 15 kg m s^{-1} and direction N35°W.

Question 4

A ball is thrown from a window with an initial upward velocity of 2.0 m s^{-1}. It hits the ground after a period of 2.0 s. Determine how high above the ground the window is.

Question 5

A 4.0 kg mass begins to move down a surface when it is raised to an angle of 10°. Calculate the maximum static friction that may be applied.

Question 6

The average $\frac{r^3}{T^2}$ value for our solar system is 3.35×10^{18} m^3 s^{-2}. The mean orbital radius of an unknown planet is noted to be 6.55×10^{10} m. Use this value, and Kepler's third law, to determine the orbital period of the unknown planet.

Question 7

Two 4.0 kg masses are connected over a frictionless pulley. One mass hangs vertically, while the other mass is on an inclined plane making an angle of 22° to the horizontal. Assume that the friction between the plane and the mass is a constant 8 N, determine the net force and the acceleration of the system.

Question 8

Voyager 1, launched on 5 September 1977, had a launch mass of 825.5 kg. Its force of weight has varied while it has travelled across our Solar System. Calculate its weight (force gravitational) when it was nearest Saturn, at a distance of 101 000 m from its centre, in 1981. Use: $M_S = 5.00 \times 10^{26}$ kg.

Question 9

A 120 kg mountain bike cyclist travels over a speed hump of radius 1.8 m at a speed of 32 kmh^{-1}. Calculate whether the cyclist leaves the track or not.

Question 10

A ball is kicked with an initial velocity of 12 m s^{-1} at an angle of 40° to a playing field that is 3 m above the initial ground level. For what period of time will the ball be airborne?

Question 11

A particle in a laboratory is noted to have an approximate lifetime of 3.0×10^{-8} s. When moving at a very high speed in the laboratory, the electron is observed to have a lifetime of 9.8×10^{-8} s. How fast was the electron moving when it was observed to have this lifetime?

Question 12

In a double-slit experiment, light with wavelength 589 nm is used to illuminate twin slits that are separated by 0.015 mm. The pattern produced is observed on a wall 1.20 m from the slits. Determine the position of:

a the first interference maximum (bright spot)

b the first interference minimum (dark spot).

Question 13

Describe one phenomenon that supports the Big Bang theory?

Question 14

If a single loop of wire with a radius of 0.060 m is placed perpendicularly within a uniform 0.050 T magnetic field, determine the emf produced if it is completely removed from the magnetic field in a period of 0.050 s.

Question 15

Determine the energy required to give a proton a velocity of 0.90*c* beginning from rest.

Question 16

In a double-slit experiment light with wavelength 609 nm is used to illuminate twin slits. The pattern produced is observed on a wall 1.80 m from the slits. The sixth interference minimum (dark spot) was found to be at a position 9.6 mm from the central bright spot. Use this data to calculate the slit separation.

Question 17

A black body is known to have a surface temperature of 3800 K. Use the value of $b = 2.898 \times 10^{-3}$ m K to determine the peak wavelength.

Question 18

Determine whether the following reaction is possible under the conservation of lepton and conservation of baryon number:

$$n \rightarrow p + e^{-} + \nu_e$$

Question 19

Two energy levels within a particular atom are 8.60 eV and an unknown, lower energy, *x* eV. When an electron of this element returns from the higher level to the lower energy level, radiation of wavelength 550 nm is emitted. Determine the value of *x*.

Question 20

Describe what a helical line represents in a Feynman diagram. On what axis is time represented?

COMBINATION-RESPONSE QUESTIONS

Question 1

Data values for several natural satellites of the Jovian system are provided. Use the data for the moon Ganymede to determine the ratio of $\frac{R^3}{T^2}$ and hence determine the missing orbital radius and orbital period values of Jupiter's moons Europa and Io.

MOON	ORBITAL PERIOD, *T* (DAYS)	ORBITAL RADIUS, *r* (m)
Io		4.21×10^8
Europa	3.56	
Ganymede	7.10	1.06×10^9

Question 2

An Atlas rocket weighing 597 000 kg is launched from ground level. When it is at an altitude of 1 km its vertical velocity is 800 m s^{-1}.

a What is the gravitational potential energy of the rocket at an altitude of 1 km?

b What is the E_k of the rocket when the rocket is at this altitude?

c How much work was done on the rocket to change its gravitational potential and kinetic energy?

9780170412643

Question 3

A 2.5 kg bucket is whirled in a vertical circle on the end of a string of length 1.0 m at a constant speed of 4.0 m s^{-1}. Calculate the tension in the string at the top and at the bottom of the circle.

Question 4

Determine the escape velocity required for a rocket to escape Mars' gravitational attraction. Use:

- $G = 6.67 \times 10^{-11}$ N m^2 kg^{-2}
- $M_M = 6.39 \times 10^{23}$ kg
- $r_M = 3.39 \times 10^6$ m

Question 5

A 1450 kg vehicle travels around a bend of radius 45 m which is banked at an angle of 11° to the horizontal at a velocity of 22 m s^{-1}. Determine the force centripetal required to allow the vehicle to complete the bend.

Question 6

A 240 W, 12 V AC supply is connected to the input terminals of a transformer. The primary coil is wound with 1500 turns. The output emf is 60.0 V. Assume there is no power loss in the transformer. Determine the number of turns on the secondary coil and state whether this is a step-up or step-down transformer.

Question 7

The reaction diagram for a Beta β^- decay is shown below. Describe the nature of this reaction, what each particle represents and how the law of conservation is applied.

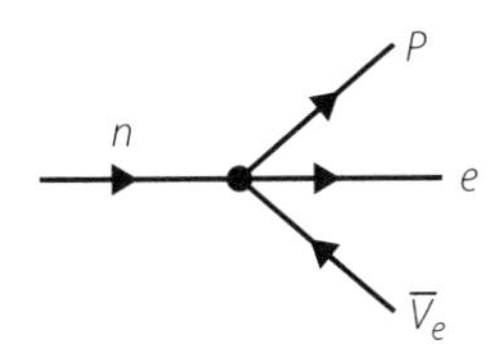

Question 8

The graph below shows the results of a photoelectric experiment using magnesium metal.

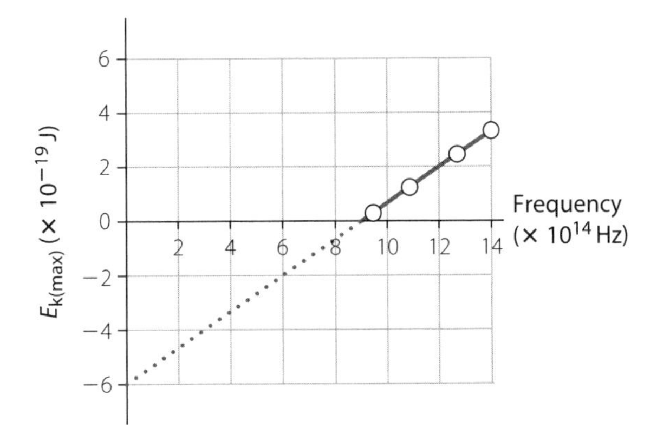

Determine a value for the work function of magnesium.

Question 9

State the four postulates that are the basis of Bohr's atomic model.

Question 10

The table below shows how electric field strength changing at increasing distances from a point source, *Q*.

R (m)	E (NC^{-1})
0.02	19.25
0.04	4.85
0.06	2.17
0.08	1.21
0.10	0.79

Construct a graph by first extending the tables and making the data linear to then determine the strength of the point source *Q* emitting this field.

9780170412643

ANSWERS

CHAPTER 1 REVISION

1.1 BEARINGS

1 a

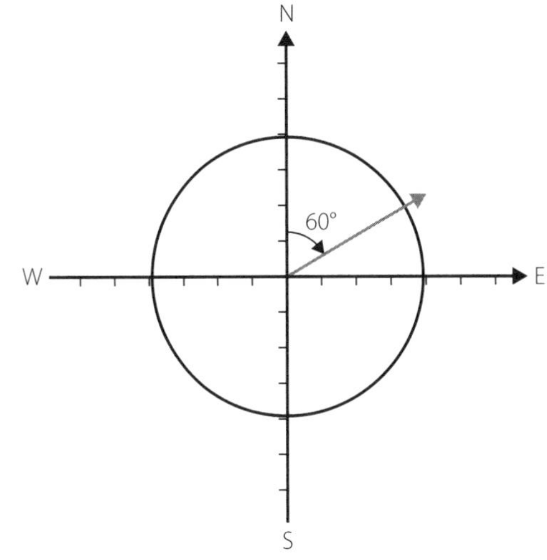

Scale = 1:10 m s^{-1}

b

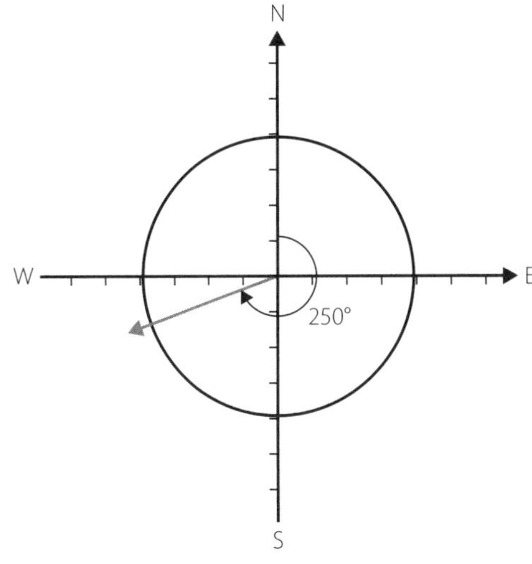

Scale = 1:10 m s^{-1}

2 a From right to left: 7 m, 35 km, 60 m and 15 km

b i From right to left: N60°, S60°E, S60°W and N30°W

ii From right to left: 60°, 120°, 240° and 330°

3 a 40°

b 330°

c 150°

d 200°

4 a N50°E

b S30°E

c S50°W

d N40°W

5 a P = 20°; Q = 150°; R = 40°; S = 110°

b $0° \leq \theta \leq 180°$

1.2 VECTORS IN TWO DIRECTIONS

1 a

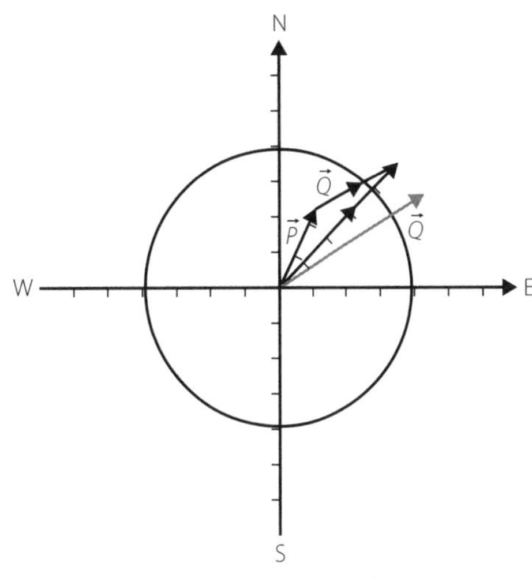

Scale = 1:10 km h^{-1}

Answer: 48 km h^{-1}, 45° true

b

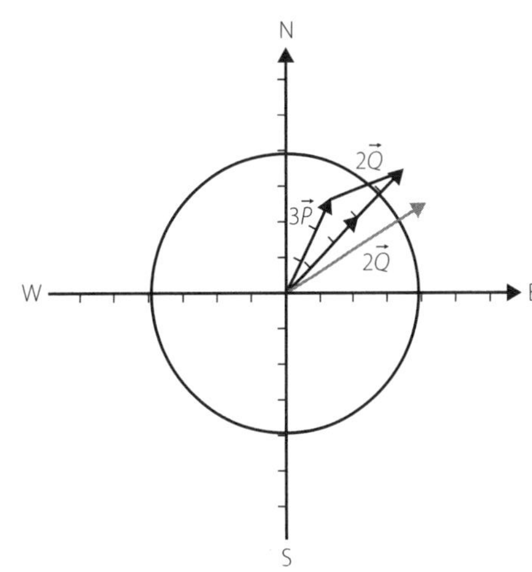

Scale = 1:25 km h^{-1}

Answer: 118 km h^{-1}, 42° true

c

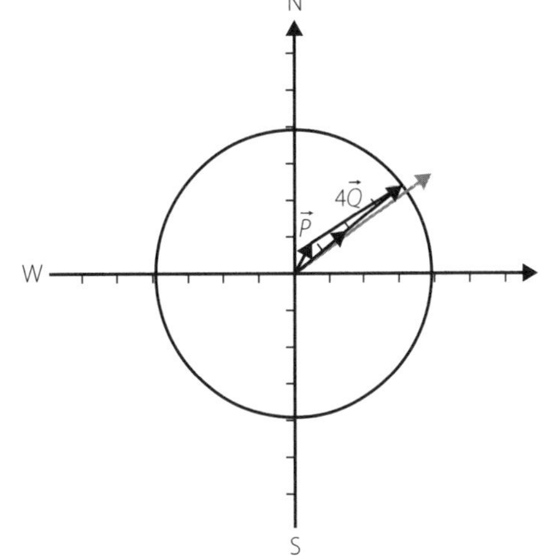

Scale = 1:25 km h^{-1}

Answer: 96 km h^{-1}, 61° true

d

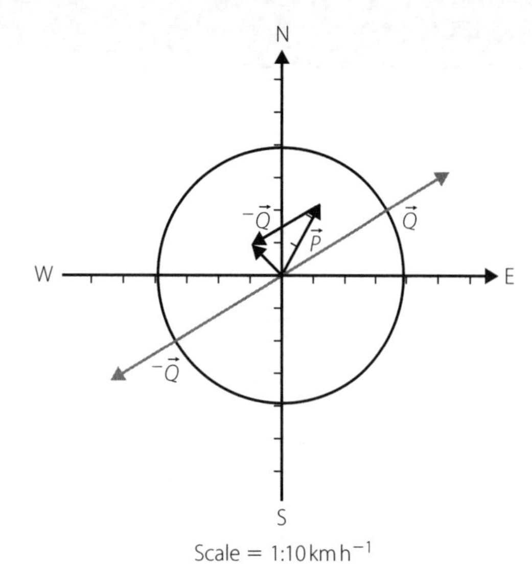

Answer: 13 km h^{-1}, 315° true

e

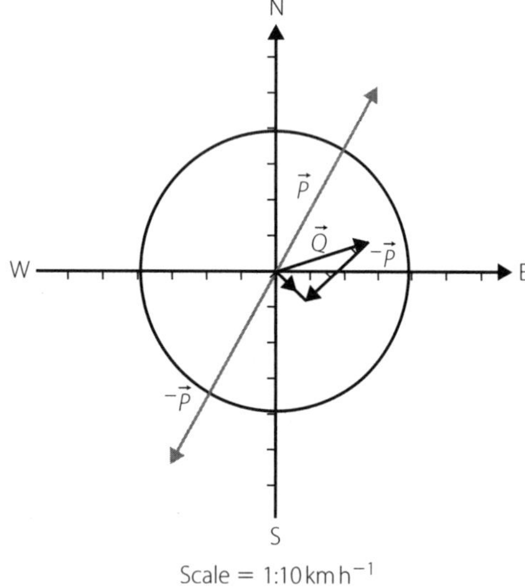

Answer: 13 km h^{-1}, 135° true

f

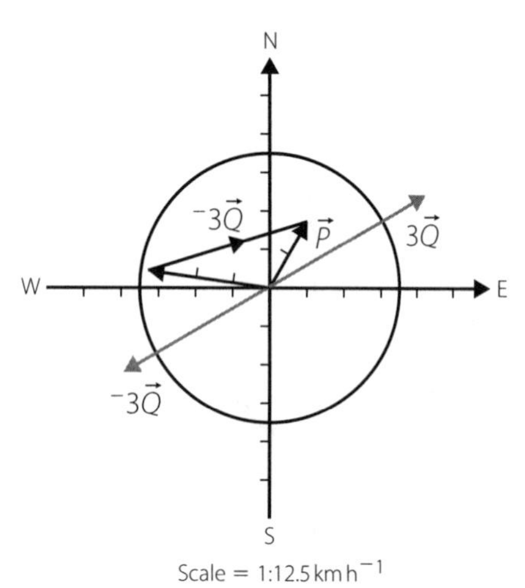

Answer: 41 km h^{-1}, 278° true

2 a

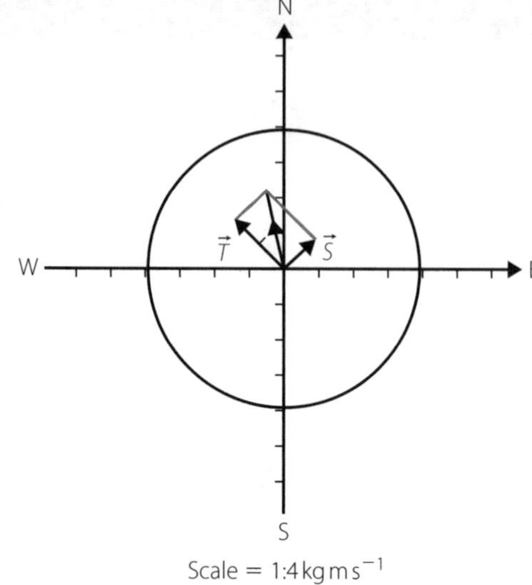

Answer: 10 kg m s^{-1}, 352° true

b

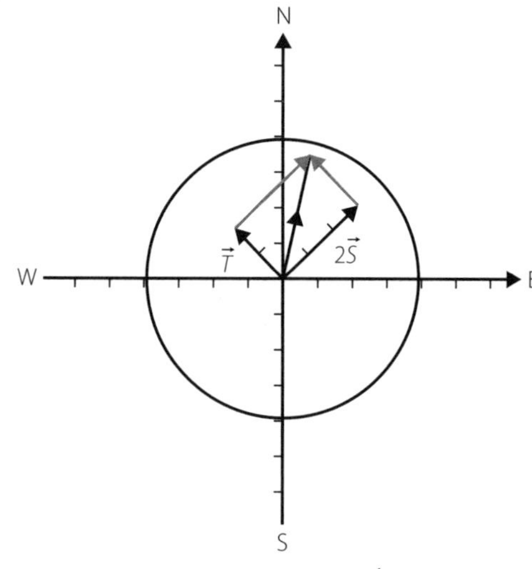

Answer: 14 kg m s^{-1}, 11° true

c

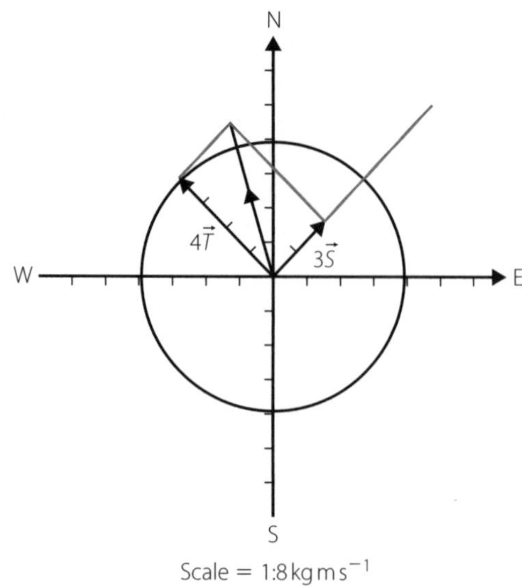

Answer: 37 kg m s^{-1}, 106° true

d

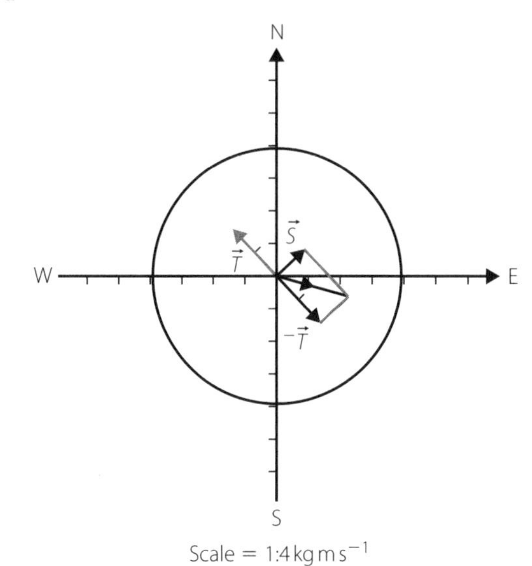

Scale = 1:4 kg m s^{-1}

Answer: 10 kg m s^{-1}, 98° true

e

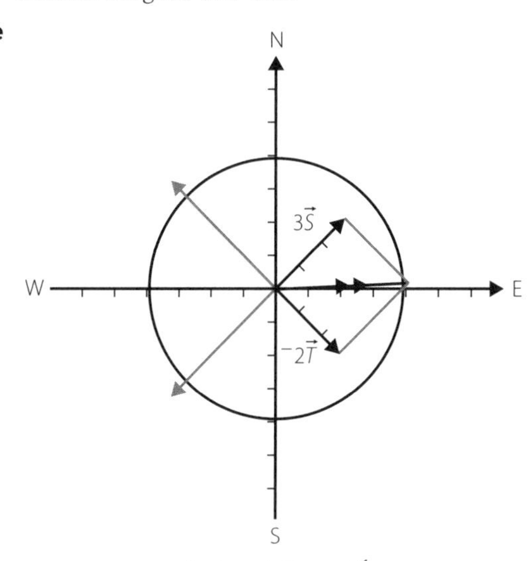

Scale = 1:6 kg m s^{-1}

Answer: 22 kg m s^{-1}, 79° true

f

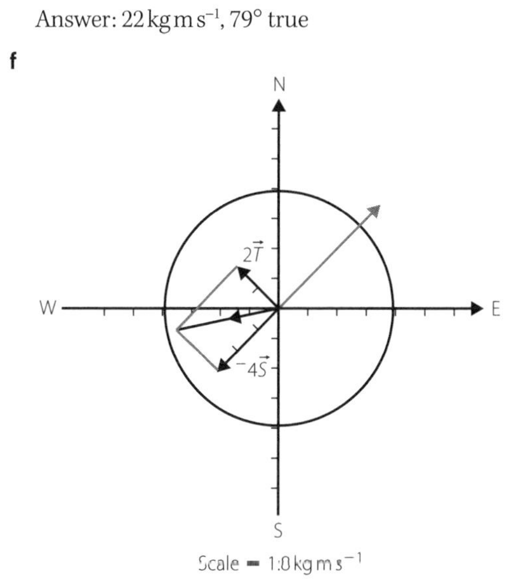

Scale = 1:0 kg m s^{-1}

Answer: 29 kg m s^{-1}, 259° true

3 a

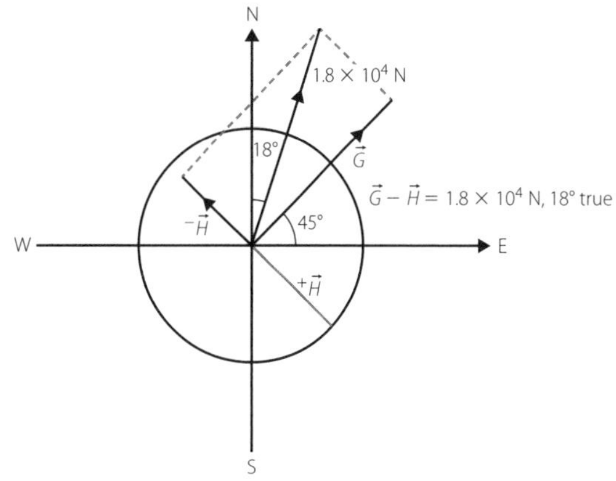

Scale = 1 division represents 8×10^3 N

b

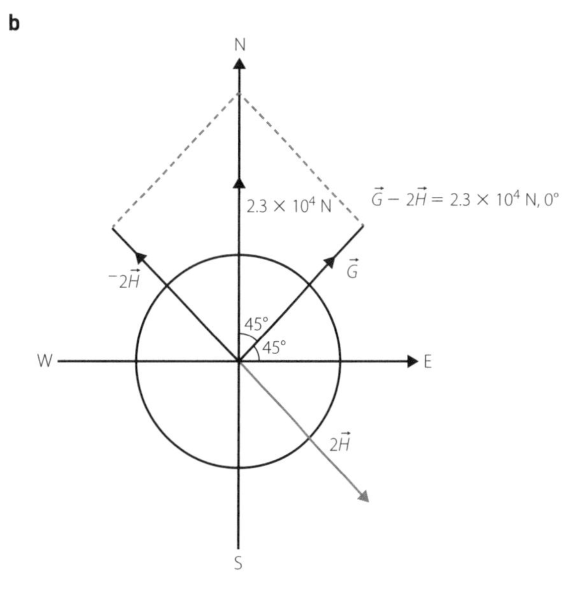

Scale = 1 division represents 8×10^3 N

c

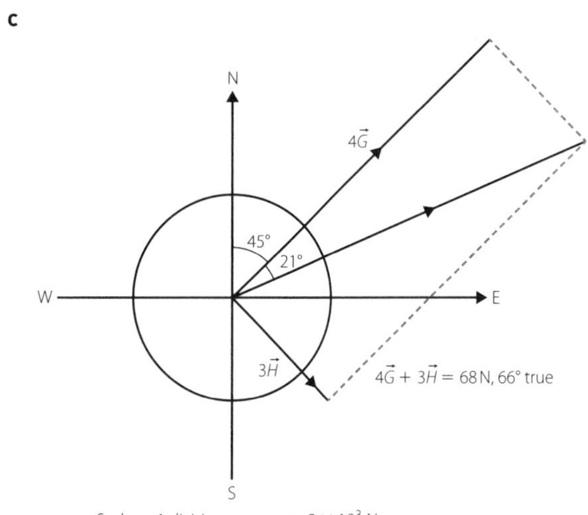

Scale = 1 division represents 8×10^3 N

d

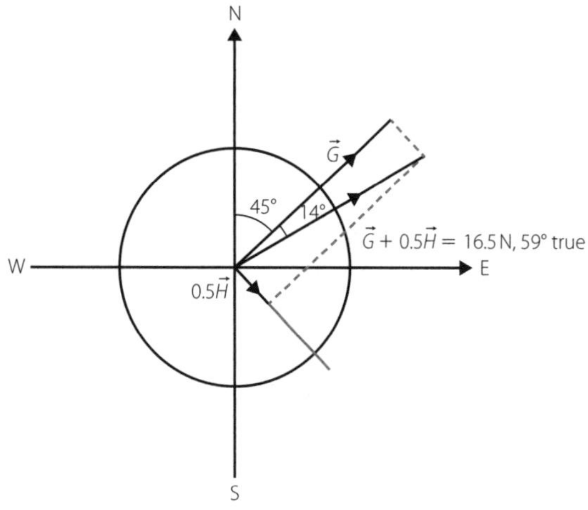

Scale = 1 division represents 8×10^3 N

e

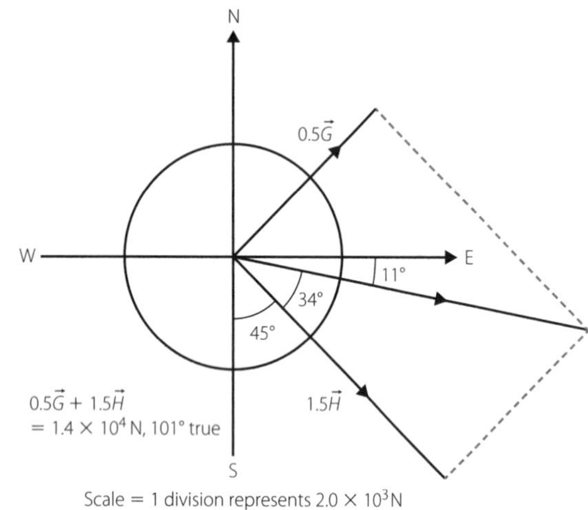

Scale = 1 division represents 2.0×10^3 N

f

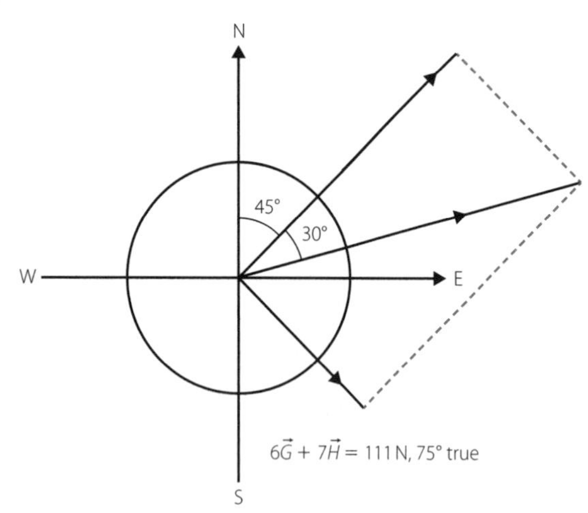

Scale = 1 division represents 8.0×10^3 N

1.3 COMPONENTS OF VECTORS

1 a

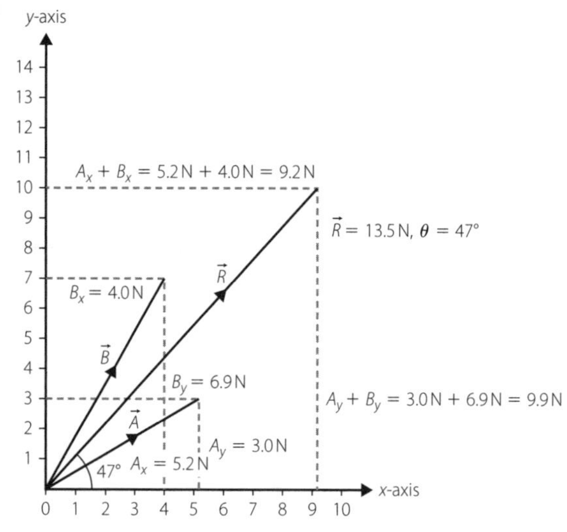

b

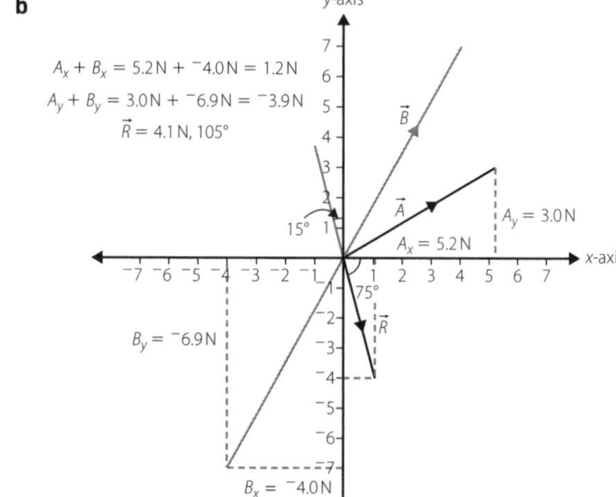

2 a

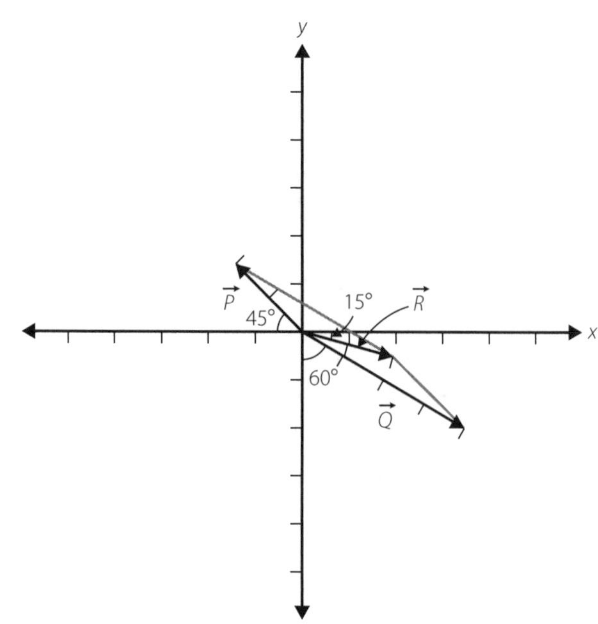

Scale: 1 division represents 1.0×10^{-4} kg m s^{-1}

Answer: 2.1×10^{-4} kg m s^{-1}, ⁻16°

9780170412643

b

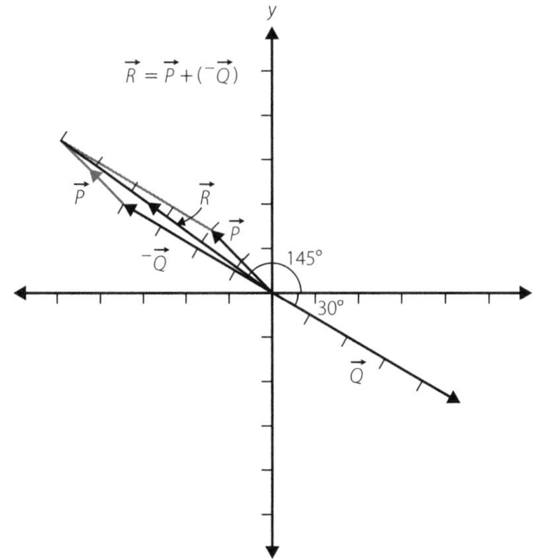

Scale: 1 division represents 1.0×10^{-4} kg m s^{-1}

Answer: 6.0×10^{-4} kg m s^{-2}

3 a

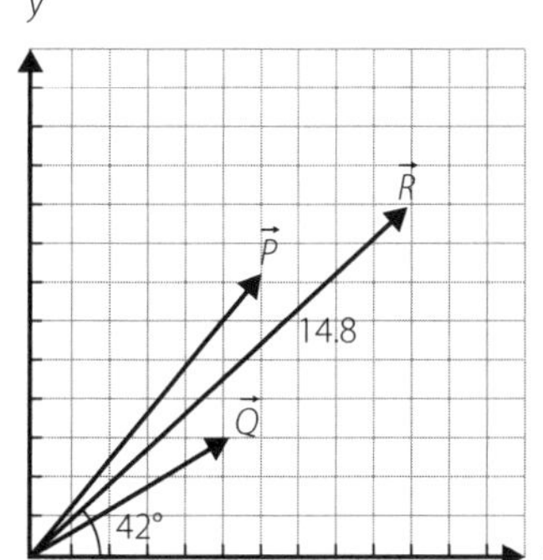

$P(6,7)$ $Q(5,3)$

$P_x + Q_x = 6 + 5 = 11$

$P_y + Q_y = 7 + 3 = 10$

$|\vec{R}| = 14.8$

$\theta = 42°$

b

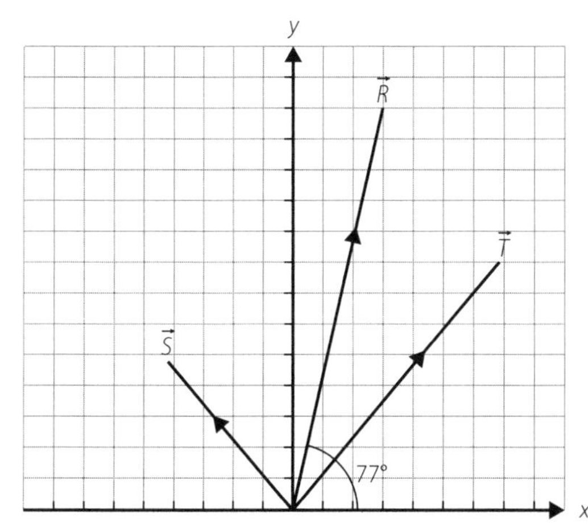

$S(^-4,5)$ $T(7,8)$

$R_x = S_x + T_x = {}^-4 + 7 = 3$

$R_y = S_y + T_y = 5 + 8 = 13$

$R = 13.3$

$\theta = 77°$

4 a 40°

b 330°

c 150°

d 200°

5 a N50°E

b S30°E

c i S50°W

ii N40°W

CHAPTER 1 EVALUATION

MULTIPLE-CHOICE

1 A

2 A

3 B

4 C

5 D

6 A

7 B

8 D

9 D

SHORT ANSWER

10 a

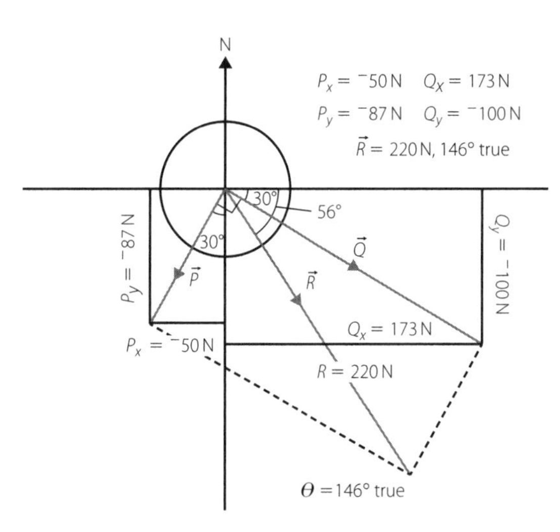

b See part **a**

c $2P_x = 2 \times {}^-50\,\text{N} = {}^-100\,\text{N}; Q_x = 173\,\text{N}$

$R_x = 100\,\text{N} + 173\,\text{N} = 273\,\text{N}$

$2P_y = 2 \times {}^-87\,\text{N} = {}^-174\,\text{N}; Q_y = {}^-100\,\text{N}$

$R_y = {}^-174\,\text{N} + {}^-100\,\text{N} = {}^-274\,\text{N}$

$R = \sqrt{(273\,\text{N})^2 + (274\,\text{N})^2}$

$R = 387\,\text{N}$

$\theta = 45°$

11 a

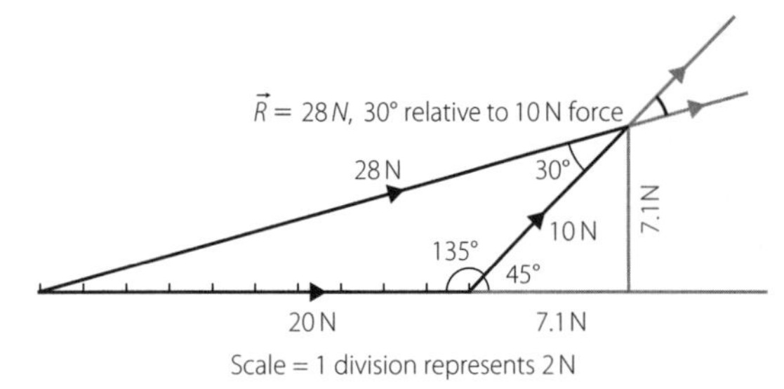

b **i** See part **a**

ii See part **a**

12 a

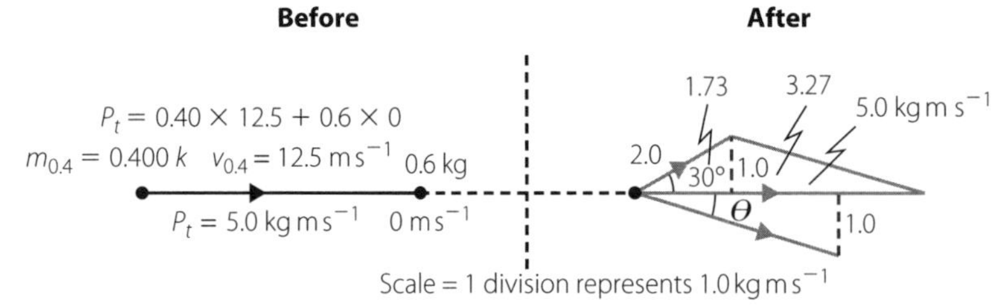

b **i** $\left(p'_{0.4}\right)_{\parallel} = p'_{0.4}\cos 30°$

$\left(p'_{0.4}\right)_{\parallel} = 2.0\,\text{kg}\,\text{m}\,\text{s}^{-1} \times \cos 30°$

$\left(p'_{0.4}\right)_{\parallel} = 1.73\,\text{kg}\,\text{m}\,\text{s}^{-1}$

ii $\left(p'_{0.6}\right)_{\parallel} = 5.0\,\text{kg}\,\text{m}\,\text{s}^{-1} - 1.73\,\text{kg}\,\text{m}\,\text{s}^{-1}$

$\left(p'_{0.6}\right)_{\parallel} = 3.27\,\text{kg}\,\text{m}\,\text{s}^{-1}$

13 a

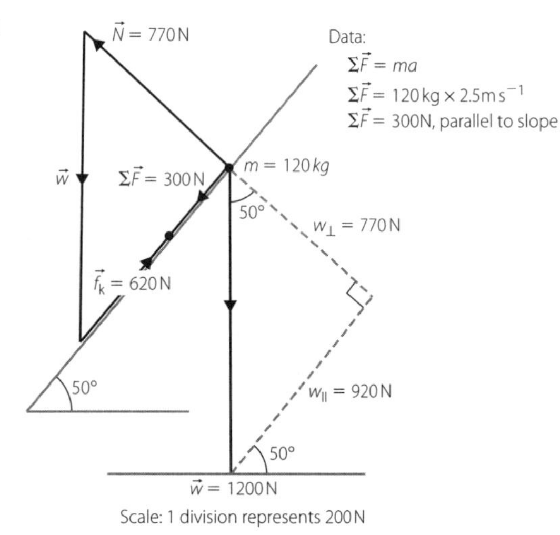

b See part **a**

CHAPTER 2 REVISION

2.1 MOTION IN THE HORIZONTAL DIRECTION

1

INITIAL LAUNCH SPEED (m s⁻¹)	LAUNCH ANGLE (°)	HORIZONTAL COMPONENT OF INITIAL LAUNCH SPEED (m s⁻¹)
10	30	8.7
2.5	60	1.3
15	40	11.5
39.6	45	28
16.4	70	5.6
72	71	24
6.3	46	4.8

2 $u_x = \dfrac{450\,\text{m}}{16\,\text{s}}$

$\Rightarrow u_x = 28.1\,\text{m}\,\text{s}^{-1}$

$u_x = u\cos\theta$

$\Rightarrow u = \dfrac{u_x}{\cos\theta}$

$\Rightarrow u = \dfrac{28.1\,\text{m}\,\text{s}^{-1}}{\cos 30°}$

$\Rightarrow u = 32\,\text{m}\,\text{s}^{-1}$

3 $u_x = \dfrac{15\,\text{m}}{2 \times 2.3\,\text{s}}$

$\Rightarrow u_x = 3.3\,\text{m}\,\text{s}^{-1}$

$u = \dfrac{u_x}{\cos\theta}$

$u = \dfrac{3.3\,\text{m}\,\text{s}^{-1}}{\cos 76}$

$u = 13.6\,\text{m}\,\text{s}^{-1}$

4 $u_x = \dfrac{55\,\text{m}}{3.5\,\text{s}}$

$\Rightarrow u_x = 15.7\,\text{m s}^{-1}$

$\theta = \cos^{-1}\left(\dfrac{15.7\,\text{m s}^{-1}}{32\,\text{m s}^{-1}}\right)$

$\theta = 61°$

2.2 MOTION IN THE VERTICAL DIRECTION

1

INITIAL LAUNCH SPEED (m s⁻¹)	LAUNCH ANGLE (°)	VERTICAL COMPONENT OF INITIAL LAUNCH SPEED (m s⁻¹)
10	30	5.0
2.5	60	2.2
15	40	9.6
39.6	45	28
6.0	70	5.6
72	19.5	24
20	13.9	4.8

2 $v_y^2 = 2gs_y + u_y^2$

$v_y = \sqrt{2gs_y}$

$v_y = v\sin\theta = \sqrt{2gs_y}$

$v = \dfrac{\sqrt{2gs_y}}{\sin\theta}$

$v = \dfrac{\sqrt{2\times 9.8\,\text{m s}^{-2}\times 75\,\text{m}}}{\sin 70°}$

$v = 41\,\text{m s}^{-1}$

2.3 ALGEBRAIC ANALYSIS OF PROJECTILE MOTION

1

INITIAL LAUNCH SPEED (m s⁻¹)	LAUNCH ANGLE (°)	HORIZONTAL COMPONENT OF INITIAL LAUNCH SPEED (m s⁻¹)	VERTICAL COMPONENT OF INITIAL LAUNCH SPEED (m s⁻¹)
40	60	20	34.6
11.5	30	10	5.8
28	40	21	18
19.2	38.7	15	12
5.1	58	2.7	4.3

2

$v = 21.0\,\text{m s}^{-1}$

$m = 6.8\,\text{g}$

28 m

a $s_y = 28.0\,\text{m}$; $u_y = 0$; $v_y = ?$; $g = {}^{+}9.8\,\text{m s}^{-2}$; $t = ?$

$s_y = \dfrac{1}{2}gt^2 + u_y t$

$\Rightarrow t = \sqrt{\dfrac{2s_y}{g}}$

$\Rightarrow t = \sqrt{\dfrac{2\times 28.0\,\text{m}}{{}^{+}9.8\,\text{m s}^{-2}}}$

$\Rightarrow t = 2.4\,\text{s}$

b $u_x = \dfrac{s_x}{t}$

$s_x = u_x t$

$s_x = 21.0\,\text{m s}^{-1}\times 2.4\,\text{s}$

$s_x = 50.2\,\text{m}$

c $s_y = 14.0\,\text{m}$; $u_y = 0\,\text{m s}^{-1}$; $v_y = ?$; $g = {}^{+}9.8\,\text{m s}^{-2}$; $t = ?$

$v_y^2 = 2gs_y + u_y^2$

$\Rightarrow v_y = \sqrt{2gs_y + u_y^2}$

$\Rightarrow v_y = \sqrt{2\times {}^{+}9.8\,\text{m s}^{-2}\times 14.0\,\text{m}}$

$\Rightarrow v_y = 16.6\,\text{m s}^{-1}$

$v_{x,14} = \sqrt{u_x^2 + u_y^2}$

$\Rightarrow v_{x,14} = \sqrt{(16.6\,\text{m s}^{-1})^2 + (21.0\,\text{m s}^{-1})^2}$

$\Rightarrow v_{x,14} = 26.7\,\text{m s}^{-1}$

$\theta = \tan^{-1}\left(\dfrac{v_y}{u_x}\right)$

$\Rightarrow \theta = \tan^{-1}\left(\dfrac{16.6\,\text{m s}^{-1}}{21.0\,\text{m s}^{-1}}\right)$

$\Rightarrow \theta = 38.3°$

3

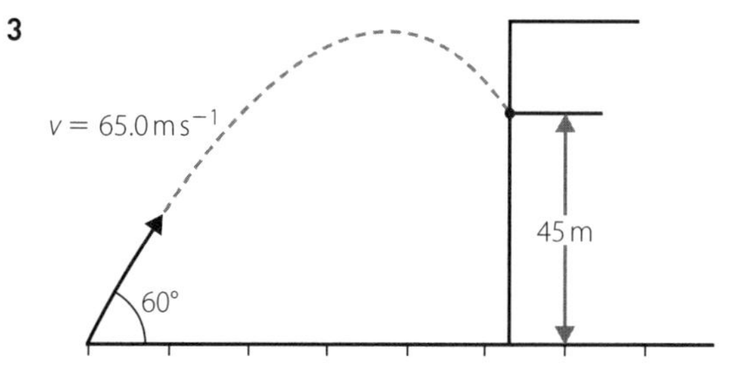

a $u_y = u\sin\theta$

$\Rightarrow u_y = 65.0\,\text{m s}^{-1}\times\sin 60°$

$\Rightarrow u_y = 56.3\,\text{m s}^{-1}$

$s_y = ?$; $u_y = 56.3\,\text{m s}^{-1}$; $v_y = 0$; $g = {}^{-}9.8\,\text{m s}^{-2}$; $t = ?$

$v_y^2 = 2gs_y + u_y^2$

$\Rightarrow s_y = \dfrac{v_y^2 - u_y^2}{2g}$

$\Rightarrow s_y = \dfrac{0^2 - (56.3\,\text{m s}^{-1})^2}{2\times {}^{-}9.8\,\text{m s}^{-2}}$

$\Rightarrow s_y = 162\,\text{m}$

b $v_y = gt + u_y$

$\Rightarrow t = \dfrac{v_y - u_y}{g}$

$\Rightarrow t = \dfrac{0 - 56.3\,\text{m s}^{-1}}{^{-}9.8\,\text{m s}^{-2}}$

$\Rightarrow t = 5.74\,\text{s}$

c $s_y = 45.0\,\text{m};\ u_y = 56.3\,\text{m s}^{-1};\ v_y = ?;\ g = {}^{-}9.8\,\text{m s}^{-2};\ t = ?$

$s_y = \frac{1}{2}gt^2 + u_y t$

$\frac{1}{2}gt^2 + u_y t - s_y = 0$

$\Rightarrow \frac{1}{2}\left({}^{-}9.8\,\text{m s}^{-2}\right)t^2 + \left(56.3\,\text{m s}^{-1}\right)t - 45.0\,\text{m} = 0$

$\Rightarrow 4.9t^2 - 56.3t - 45 = 0$

$\Rightarrow t = \dfrac{-\left({}^{-}56.3\right) \pm \sqrt{(56.3)^2 - 4 \times 4.9 \times {}^{-}45}}{2 \times 4.9}$

$\Rightarrow t = 5.74 + 6.49$

$\Rightarrow t = 12.2\,\text{s}$

d $u_x = \dfrac{s_x}{t}$

$s_x = u_x t$

$s_x = 32.5\,\text{m s}^{-1} \times 12.2\,\text{s}$

$s_x = 398\,\text{m}$

4

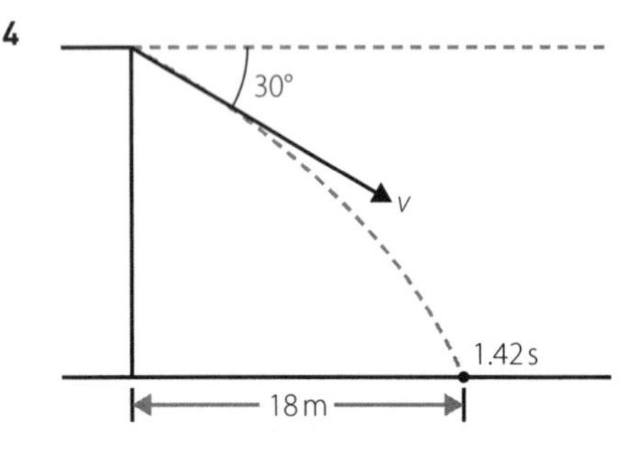

a $u_x = u\cos\theta$

$\Rightarrow u_x = u\cos 30°$

$\Rightarrow u_x = u\dfrac{\sqrt{3}}{2} = 0.866 \times u$

$u_x = \dfrac{s_x}{t} = \dfrac{18.0\,\text{m}}{1.42\,\text{s}} = 12.67\,\text{m s}^{-1}$

$\Rightarrow 0.866 \times u = 12.68\,\text{m s}^{-1}$

$\Rightarrow u = \dfrac{12.68\,\text{m s}^{-1}}{0.866}$

$\Rightarrow u = 14.64\,\text{m s}^{-1}$

b $u_x = 12.68\,\text{m s}^{-1};\ v_y = 8.55\,\text{m s}^{-1}$

$v_{(18,25)} = \sqrt{u_x^2 + v_y^2}$

$\Rightarrow v_{(18,25)} = \sqrt{\left(12.68\,\text{m s}^{-1}\right)^2 + \left(8.55\,\text{m s}^{-1}\right)^2}$

$\Rightarrow v_{(18,25)} = 15.3\,\text{m s}^{-1}$

5

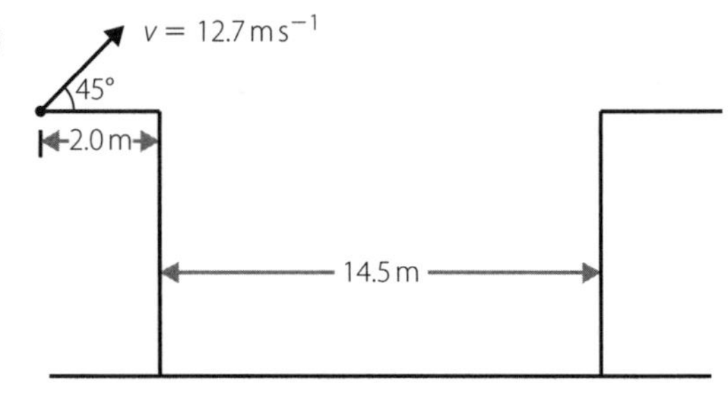

$R = \dfrac{u^2 \sin 2\theta}{g}$

$R = \dfrac{\left(12.7\,\text{m s}^{-1}\right)^2 \times \sin(2 \times 45°)}{9.8\,\text{m s}^{-2}}$

$R = 16.5\,\text{m}$

The ball just reaches the other building.

2.4 SOLUTION STRATEGY: PROJECTILE MOTION

1 $s_y = 15\,\text{m};\ u_y = 0\,\text{m s}^{-1};\ v_y = ?;\ g = {}^{+}9.8\,\text{m s}^{-2};\ t = ?$

$v_y^2 = 2gs + u_y^2$

$\Rightarrow v_y = \sqrt{2gs + u_y^2}$

$\Rightarrow v_y = \sqrt{2 \times {}^{+}9.8\,\text{m s}^{-2} \times 15\,\text{m} + \left(0\,\text{m s}^{-1}\right)^2}$

$\Rightarrow v_y = 17.1\,\text{m s}^{-1}$

$v_{x,y} = \sqrt{u_x^2 + v_y^2}$

$\Rightarrow v_{x,y} = \sqrt{\left(9.0\,\text{m s}^{-1}\right)^2 + \left(17.1\,\text{m s}^{-1}\right)^2}$

$\Rightarrow v_{x,y} = 19.4\,\text{m s}^{-1}$

2 a $u_{x,\text{W}} = 13\,\text{m s}^{-1} \times \cos 60°$

$\Rightarrow u_{x,\text{W}} = 6.5\,\text{m s}^{-1}$

b By symmetry, the vertical component at Y is the same magnitude, but opposite direction, to that at X:

$u_{y,\text{W}} = u_{y,\text{Y}} = 13\,\text{m s}^{-1} \times \sin 60°$

$\Rightarrow u_{y,\text{Y}} = 2.9\,\text{m s}^{-1}$

c $s_y = ?;\ u_{y,\text{W}} = 11\,\text{m s}^{-1};\ v_{y,\text{X}} = 0\,\text{m s}^{-1};\ g = {}^{-}9.8\,\text{m s}^{-2};\ t_{\text{WX}} = ?$

$v_y = gt + u_y$

$\Rightarrow t_{\text{WX}} = \dfrac{v_y - u_y}{g}$

$\Rightarrow t_{\text{WX}} = \dfrac{0\,\text{m s}^{-1} - 11\,\text{m s}^{-1}}{^{-}9.8\,\text{m s}^{-2}}$

$\Rightarrow t_{\text{WX}} = 1.1\,\text{s}$

$t_{\text{WY}} = 2t_{\text{WX}}$

$\Rightarrow t_{\text{WY}} = 2.2\,\text{s}$

d $s_{\text{XW}} = ?;\ u_{y,\text{W}} = 11\,\text{m s}^{-1};\ v_{y,\text{X}} = 0\,\text{m s}^{-1};\ g = {}^{-}9.8\,\text{m s}^{-2};\ t_{\text{WX}} = 1.1\,\text{s}$

$s_y = \dfrac{u_y + v_y}{2}t$

$s_{\text{XW}} = \dfrac{11\,\text{m s}^{-1} + 0\,\text{m s}^{-1}}{2} \times 1.1\,\text{s}$

$s_{\text{XW}} = 6.5\,\text{m}$

9780170412643

e $s_{WY} = ?;\ u = 13\,m\,s^{-1};\ \theta = 60°;\ g = {}^{-}9.8\,m\,s^{-2}$

$$R = \frac{u^2 \sin 2\theta}{g}$$

$$\Rightarrow s_{WY} = \frac{(13\,m\,s^{-1})^2 \times \sin(2\times 60°)}{9.8\,m\,s^{-2}}$$

$$\Rightarrow s_{WY} = 14.9\,m$$

f $s_{y,Z} = {}^{-}1.5\,m;\ u_{y,W} = 11\,m\,s^{-1};\ v_{y,Z} = ?;\ g = {}^{-}9.8\,m\,s^{-2};\ t_{WY} = ?$

$$v_y^2 = 2gs_y + u_y^2$$

$$\Rightarrow v_{y,Z} = \sqrt{2gs_{y,Z} + u_y^2}$$

$$\Rightarrow v_{y,Z} = \sqrt{2\times {}^{-}9.8\,m\,s^{-2} \times ({}^{-}1.5\,m) + (11\,m\,s^{-1})^2}$$

$$\Rightarrow v_{y,Z} = 12.5\,m\,s^{-1}$$

g $v_{y,Z} = \sqrt{u_{x,W}^2 + v_{y,Z}^2}$

$$\Rightarrow v_{y,Z} = \sqrt{(6.5\,m\,s^{-1})^2 + (12.5\,m\,s^{-1})^2}$$

$$\Rightarrow v_{y,Z} = 14\,m\,s^{-1}$$

h $\theta = \tan^{-1}\left(\frac{v_{y,Z}}{u_{x,\frac{W}{Z}}}\right)$

$$\theta = \tan^{-1}\left(\frac{14\,m\,s^{-1}}{6.5\,m\,s^{-1}}\right)$$

$$\theta = 65°$$

i $s_{y,Z} = {}^{-}1.5\,m;\ u_{y,W} = 11\,m\,s^{-1};$

$v_{y,Z} = ?;\ g = {}^{-}9.8\,m\,s^{-2};\ t_{WY} = ?$

$$s_y = \frac{1}{2}gt^2 + u_y t$$

$$\Rightarrow {}^{-}1.5\,m = \frac{1}{2}\times({}^{-}9.8\,m\,s^{-2})t^2 + 11\,m\,s^{-1} \times t$$

$$\Rightarrow 4.9t^2 - 11t - 1.5 = 0$$

$$\Rightarrow t^2 - 2.29t - 0.31 = 0$$

$$\Rightarrow t = \frac{-({}^{-}2.29) \pm \sqrt{({}^{-}2.29)^2 - 4\times 1\times {}^{-}0.31}}{2\times 1}$$

$$\Rightarrow t = 2.42\,s \text{ or } t = {}^{-}0.13\,s$$

$$\Rightarrow t = 2.42\,s$$

j $s_x = ?;\ u_x = 6.5\,m\,s^{-1};\ v_x = 6.5\,m\,s^{-1};\ a_x = 0\,m\,s^{-2};\ t = 2.42\,s$

$$v_x = \frac{s_x}{t}$$

$$\Rightarrow s_x = v_x t$$

$$\Rightarrow s_x = 6.5\,m\,s^{-1} \times 2.42\,s$$

$$\Rightarrow s_x = 16\,m$$

3 a $R = 2.0\,m;\ u = 4.8\,m\,s^{-1};\ \theta = 23°;\ g = {}^{-}9.8\,m\,s^{-2}$

$$R = \frac{u^2 \sin 2\theta}{g}$$

$$\Rightarrow R = \frac{(4.8\,m\,s^{-1})^2 \times \sin(2\times 23°)}{9.8\,m\,s^{-2}}$$

$$\Rightarrow R = 1.7\,m$$

b $u_x = u\cos\theta \Rightarrow u_x = 4.9\,m\,s^{-1} \times \cos 23° = 4.5\,m\,s^{-1}$

$u_y = u\sin\theta \Rightarrow u_y = 4.9\,m\,s^{-1} \times \sin 23° = 1.9\,m\,s^{-1}$

$s_x = 2.0\,m;\ u_x = 4.5\,m\,s^{-1};\ v_x = 4.5\,m\,s^{-1};\ a_x = 0\,m\,s^{-2};\ t = ?$

$$v_x = \frac{s_x}{t}$$

$$t = \frac{s_x}{v_x} = \frac{2.0\,m}{4.5\,m\,s^{-1}} = 0.44\,s$$

$s_y = ?;\ u_y = 1.9\,m\,s^{-1};\ v_y = ?;\ a_x = {}^{-}9.8\,m\,s^{-2};\ t = 0.44\,s$

$$\Rightarrow s_y = \frac{1}{2}gt^2 + u_y t$$

$$\Rightarrow s_y = \frac{1}{2}\times({}^{-}9.8\,m\,s^{-2})\times(0.44\,s)^2 + (1.9\,m\,s^{-1})\times 0.44\,s$$

$$\Rightarrow s_y = 0.12\,m$$

4 $s_x = R = 30.0\,m;\ u = 8.0\,m\,s^{-1};\ \theta = ?;\ g = {}^{-}9.8\,m\,s^{-2}$

$$R = \frac{u^2 \sin 2\theta}{g}$$

$$\Rightarrow \sin 2\theta = \frac{gR}{u^2}$$

$$\Rightarrow 2\theta = \sin^{-1}\left(\frac{gR}{u^2}\right)$$

$$\Rightarrow \theta = \frac{1}{2}\sin^{-1}\left(\frac{gR}{u^2}\right)$$

$$\Rightarrow \theta = \frac{1}{2}\sin^{-1}\left(\frac{9.8\,m\,s^{-2} \times 30.0\,m}{(8.0\,m\,s^{-1})^2}\right)$$

$$\Rightarrow \theta = 47°$$

CHAPTER 2 EVALUATION

MULTIPLE-CHOICE

1 B
2 C
3 A
4 D
5 D
6 D
7 C
8 B
9 C
10 A
11 A
12 B
13 C
14 C
15 A
16 B
17 C
18 C
19 B
20 D

SHORT ANSWER

21 a $R = 40\,\text{m};\ u = ?;\ \theta = 50°;\ g = {}^{-}9.8\,\text{ms}^{-2}$

$$u = \sqrt{\frac{Rg}{\sin 2\theta}}$$

$$u = \sqrt{\frac{40\,\text{m} \times 9.8\,\text{ms}^{-2}}{\sin(2 \times 50°)}}$$

$$u = 20\,\text{ms}^{-1}$$

b $u_x = \frac{s_x}{t} = u\cos\theta$

$$t = \frac{s_x}{u\cos\theta}$$

$$t = \frac{40\,\text{m}}{20\,\text{ms}^{-1} \times \cos 50°}$$

$$t = 3.1\,\text{s}$$

c $s_y = 1.0\,\text{m};\ u_y = 20\,\text{ms}^{-1} \times \sin 50° = 15.3\,\text{ms}^{-1};$

$v_y = ?;\ g = {}^{-}9.8\,\text{ms}^{-2};\ t = ?$

$$s_y = \frac{1}{2}gt^2 + u_y t$$

$$1.0\,\text{m} = \frac{1}{2} \times {}^{-}9.8\,\text{ms}^{-2} \times t^2 + 15.3\,\text{ms}^{-1} \times t$$

$$\Rightarrow 4.9\,t^2 - 15.3t + 1.0 = 0$$

$$\Rightarrow t = \frac{-({}^{-}15.3) \pm \sqrt{({}^{-}15.3)^2 - 4 \times 4.9 \times 1.5}}{2 \times 4.9}$$

$$\Rightarrow t = 3.0\,\text{s}$$

d $s_y = 1.5\,\text{m};\ u_y = 15.3\,\text{ms}^{-1};\ v_y = ?$

$g = {}^{-}9.8\,\text{ms}^{-2};\ t = 3.1\,\text{s} - 3.0\,\text{s} = 0.10\,\text{s}$

$$v_y = gt + u_y$$

$$v_y = {}^{-}9.8\,\text{ms}^{-2} \times 0.10\,\text{s} + 15.3\,\text{ms}^{-1}$$

$$v_y = 14.3\,\text{ms}^{-1}$$

$$v_x = 20\,\text{ms}^{-1} \times \cos 50° = 12.9\,\text{ms}^{-1}$$

$$v = \sqrt{(12.9\,\text{ms}^{-1})^2 + (14.3\,\text{ms}^{-1})^2}$$

$$v = 19.2\,\text{ms}^{-1}$$

22 a $v = 11.0\,\text{ms}^{-1} - 3.0\,\text{ms}^{-1} = 8.0\,\text{ms}^{-1}$

b $s_y = 200\,\text{m};\ u_y = 0;\ v_y = ?;\ g = {}^{+}9.8\,\text{ms}^{-2};\ t = ?$

$$s_y = \frac{1}{2}gt^2 + u_y t$$

$$200\,\text{m} = \frac{1}{2} \times {}^{+}9.8\,\text{ms}^{-2} \times t^2$$

$$t = \sqrt{\frac{200 \times 1.0\,\text{m}}{9.8\,\text{ms}^{-2}}}$$

$$t = 4.5\,\text{ms}^{-1}$$

c $s_x = v_x t$

$$s_x = 8.0\,\text{ms}^{-1} \times 4.5\,\text{s}$$

$$s_x = 3.6\,\text{m}$$

d $s_x = v_x t$

$$s_x = 8.0\,\text{ms}^{-1} \times 4.5\,\text{s}$$

$$s_x = 3.6\,\text{m}$$

$s_y = 200\,\text{m};\ u_y = 0\,\text{ms}^{-1};\ v_y = ?;\ g = {}^{+}9.8\,\text{ms}^{-2};\ t = 4.5\,\text{s}$

$$v_y = gt + u_y$$

$$v_y = {}^{+}9.8\,\text{ms}^{-2} \times 4.5\,\text{s}$$

$$v_y = 44\,\text{m}$$

23 a $s_y = {}^{-}40\,\text{m};\ u_y = 8.0\,\text{ms}^{-1} \times \sin 30° = 4.0\,\text{ms}^{-1};$

$v_y = ?;\ g = {}^{-}9.8\,\text{ms}^{-2};\ t = ?$

$$s_y = \frac{1}{2}gt^2 + u_y t$$

$${}^{-}40\,\text{m} = \frac{1}{2} \times {}^{-}9.8\,\text{ms}^{-2} \times t^2 + 4.0\,\text{ms}^{-1} \times t$$

$$\Rightarrow 4.9\,t^2 - 4.0t - 40 = 0$$

$$\Rightarrow t = \frac{-({}^{-}4.0) \pm \sqrt{({}^{-}4.0)^2 - 4 \times 4.9 \times {}^{-}40}}{2 \times 4.9}$$

$$\Rightarrow t = 5.6\,\text{s}$$

b $s_x = v_x t$

$$s_x = 8.0\,\text{ms}^{-1} \times \cos 30° \times 5.6\,\text{s}$$

$$s_x = 39\,\text{m}$$

c

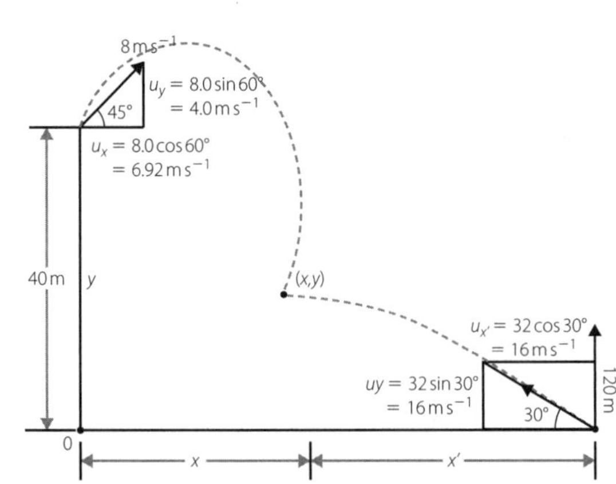

Horizontally:

Ball:	Rocket:
$x = v_x t$	$x' = v_{x'} t$
$x = 6.92t$	$x' = 16t$

$$x + x' = 120\,\text{m}$$

$$6.92t + 16t = 120$$

$$22.92t = 120$$

$$t = \frac{120}{22.92}$$

$$t = 5.23\,\text{s}$$

Ball:	Rocket:
$x = 6.92t$	$x' = 16t$
$x = 6.92\,\text{ms}^{-1} \times 5.23\,\text{s}$	$x' = 6.92\,\text{ms}^{-1} \times 5.23\,\text{s}$
$x = 36.3\,\text{m}$	$x' = 83.7\,\text{m}$

Vertically:

Rocket:

$s_y = y;\ u_y = 27.7\,\text{ms}^{-1};\ v_y = ?;\ g = {}^{-}9.8\,\text{ms}^{-2};\ t = 5.23\,\text{s}$

$s_y = \frac{1}{2}gt^2 + u_y t$

$y = \frac{1}{2} \times {}^{-}9.8\,\text{ms}^{-2} \times (5.23\,\text{s})^2 + 27.7\,\text{ms}^{-1} \times 5.23\,\text{s}$

$y = 11.0\,\text{m}$

Coordinates:

$(x, y) = (36.3\,\text{m}, 11.0\,\text{m})$

24 a $R = \frac{u^2 \sin 2\theta}{g}$

$\Rightarrow R = \frac{(36.0\,\text{ms}^{-1})^2 \times \sin(2 \times 70°)}{9.8\,\text{ms}^{-2}}$

$\Rightarrow R = 85\,\text{m}$

b $u_x = u\cos\theta = \frac{s_x}{t}$

$t = \frac{s_x}{u\cos\theta}$

$t = \frac{85\,\text{m}}{36\,\text{ms}^{-1} \times \cos 70°}$

$t = 6.9\,\text{s}$

c $h = \frac{v^2}{2g}$

$h = \frac{(36\,\text{ms}^{-1})^2}{2 \times 9.8\,\text{ms}^{-2}}$

$h = 66\,\text{m}$

d $s_y = 50\,\text{m};\ u_y = 36\,\text{ms}^{-1} \times \sin 70° = 33.8\,\text{ms}^{-1};$

$v_y = ?;\ g = {}^{-}9.8\,\text{ms}^{-2};\ t = ?$

$s_y = \frac{1}{2}gt^2 + u_y t$

$50\,\text{m} = \frac{1}{2} \times {}^{-}9.8\,\text{ms}^{-2} \times t^2 + 33.8\,\text{ms}^{-1} \times t$

$4.9\,t^2 - 33.8t + 50 = 0$

$t = \frac{-({}^{-}33.8) \pm \sqrt{({}^{-}33.8)^2 - 4 \times 4.9 \times 50}}{2 \times 4.9}$

$t = \frac{(33.8 \pm 12.8)}{9.8}\,\text{s}$

$t = 2.14\,\text{s}$ and $t = 4.76\,\text{s}$

$\Delta t = 2.62\,\text{s}$

CHAPTER 3 REVISION

3.1 RESOLVING FORCES

1 $\Sigma F_{\parallel} = ma$

$\Rightarrow \Sigma F_{\parallel} = 75\,\text{kg} \times 1.8\,\text{ms}^{-2}$

$\Rightarrow \Sigma F_{\parallel} = 135\,\text{N}$

2 a i $w = mg$

$\Rightarrow w = 5.0\,\text{kg} \times 9.8\,\text{ms}^{-2}$

$\Rightarrow w = 40\,\text{N}$

ii $w_{\parallel} = w\sin\theta$

$\Rightarrow w_{\parallel} = 40\,\text{N} \times \sin 30°$

$\Rightarrow w_{\parallel} = 20\,\text{N}$

iii $w_{\perp} = w\cos\theta$

$\Rightarrow w_{\perp} = 40\,\text{N} \times \cos 30°$

$\Rightarrow w_{\perp} = 34.6\,\text{N}$

b Perpendicular to the slope:

$\Sigma F_{\perp} = N - w_{\perp} = 0$ (Newton's second law)

$\Rightarrow N = w_{\perp}$

$\Rightarrow N$ and $w_{\perp}$ are equal and opposite but act on the same thing (the mass). They are not of the same type (N is electrostatic; $w_{\perp}$ is gravitational).

3

ANGLE (°)	MASS (kg)	WEIGHT (N)	$w_{\parallel}$ (N)	$w_{\perp}$ (N)	NORMAL FORCE (N)
20	10	98	33.5	92.1	92.1
30	15	147	73.5	127.3	127.3
23.3	20	196	77.5	180	180
60	25.5	250	216.5	125	125
37	80	784	472	626	626

4 $w_{\parallel} = mg\sin\theta = ma$

$\Rightarrow a_{\parallel} = g\sin\theta$

$\Rightarrow a_{\parallel} = 9.8\,\text{ms}^{-2} \times \sin 8.6°$

$\Rightarrow a_{\parallel} = 1.5\,\text{ms}^{-2}$

3.2 FRICTION AND MOTION ON AN INCLINED PLANE

1 a $F_{\parallel} - f_k = ma$

$\Rightarrow f = mg\sin\theta - ma$

$\Rightarrow f = 0.325\,\text{kg} \times 9.8\,\text{ms}^{-2} \times \sin 15°$

$\Rightarrow f_s = 0.82\,\text{N}$

b $F_{\parallel} - f_k = ma$

$\Rightarrow f_k = mg\sin\theta - ma$

$\Rightarrow f_k = 0.82\,\text{N} - 0.325\,\text{kg} \times 0.84\,\text{ms}^{-2}$

$\Rightarrow f_k = 0.55\,\text{N}$

c $N - mg\cos\theta = 0$

$\Rightarrow N = mg\cos\theta$

$\Rightarrow N = 0.325\,\text{kg} \times 9.8\,\text{ms}^{-2} \times \cos 15°$

$\Rightarrow N = 3.1\,\text{N}$

2 a $F_{\parallel} - f_k = ma$

$\Rightarrow f_k = mg\sin\theta - ma$

$\Rightarrow \sin\theta = \frac{f_k + ma}{mg}$

$\Rightarrow \sin\theta = \frac{4.0\,\text{N} + 0.56\,\text{kg} \times 2.3\,\text{ms}^{-2}}{2.3\,\text{kg} \times 9.8\,\text{ms}^{-2}}$

$\Rightarrow \theta = 13.6°$

b $v = u + at$

$\Rightarrow v = 0\,\text{m s}^{-1} + 2.3\,\text{m s}^{-2} \times 2.3\,\text{s}$

$\Rightarrow v = 7.4\,\text{m s}^{-1}$

3 $f_s = mg\sin\theta$

$mg\sin\theta - f_k = ma$

$\Rightarrow mg\sin\theta - 0.25 \times mg\sin\theta = ma$

$\Rightarrow 0.75 \times mg\sin\theta = ma$

$\Rightarrow a = 0.75 \times g\sin\theta$

$\Rightarrow a = 0.75 \times 9.8\,\text{m s}^{-2} \times \sin 30°$

$\Rightarrow a = 3.7\,\text{m s}^{-2}$

4

$\vec{N}$ $\quad \vec{f}_k = 0.40\vec{N}$

$m = 60\,\text{kg}$ $\quad \vec{F}(\text{by person}) = 50\,\text{N}$

$40°$ $\quad \vec{w} = 60\,\text{kg} \times 9.8\,\text{m s}^{-2}$

a $w_{\parallel} = mg\sin\theta$

$\Rightarrow w_{\parallel} = 60\,\text{kg} \times 9.8\,\text{m s}^{-2} \times \sin 40°$

$\Rightarrow w_{\parallel} = 378\,\text{N}$

b $N - [F(\text{by person}) + mg\cos\theta] = 0$

$\Rightarrow N = 50\,\text{N} + 60\,\text{kg} \times 9.8\,\text{m s}^{-2} \times \cos 40°$

$\Rightarrow N = 500\,\text{N}$

c $\Sigma F = w_{\parallel} - f_k = ma$

$\Rightarrow a = \dfrac{w_{\parallel} - f_k}{m}$

but $f_k = 0.40\,\text{N}$

$f_k = 0.40 \times 500\,\text{N}$

$f_k = 200\,\text{N}$

$\Rightarrow a = \dfrac{378\,\text{N} - 200\,\text{N}}{60\,\text{kg}}$

$\Rightarrow a = 3.0\,\text{m s}^{-2}$

5 a Static friction rises linearly from zero to a maximum at 12.2 N because the component of the weight force parallel to the slope is unable to free the surface of the box from the electrostatic and mechanical forces holding the box in place. Then, the kinetic friction is a constant 8.6 N because the forces holding the box and surface have been reduced when the surfaces can slide freely over each other.

b $f_s = mg\sin\theta$

$\Rightarrow \sin\theta = \dfrac{f_s}{mg}$

$\Rightarrow \theta = \sin^{-1}\left(\dfrac{12.2\,\text{N}}{4.5\,\text{kg} \times 9.8\,\text{m s}^{-2}}\right)$

$\Rightarrow \theta = 16°$

c i Zero; the slope is not high enough

ii $\Sigma F = F_{\parallel} - f_k = ma$

$\Rightarrow a = g\sin\theta - \dfrac{f_k}{m}$

$\Rightarrow a = 9.8\,\text{m s}^{-2} \times \sin 50° - \dfrac{8.6\,\text{N}}{4.5\,\text{kg}}$

$\Rightarrow a = 5.6\,\text{m s}^{-2}$

3.3 SOLUTION PROCEDURE: INCLINED PLANE

1

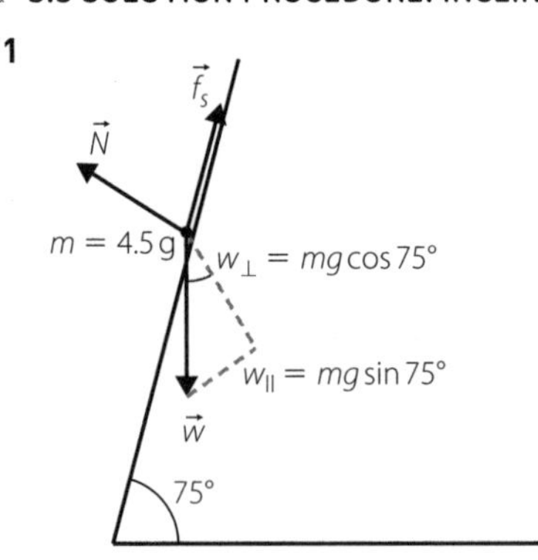

a i $w_{\parallel} = mg\sin\theta$

$\Rightarrow w_{\parallel} = 4.5 \times 10^{-3}\,\text{kg} \times 9.8\,\text{m s}^{-2} \times \sin 75°$

$\Rightarrow w_{\parallel} = 0.43\,\text{N}$

ii $w_{\perp} = mg\cos\theta$

$\Rightarrow w_{\perp} = 4.5 \times 10^{-3}\,\text{kg} \times 9.8\,\text{m s}^{-2} \times \cos 75°$

$\Rightarrow w_{\perp} = 0.11\,\text{N}$

iii $0.11\,\text{N}\ (N = W_p)$

b $f_k = 0.90 f_s = 0.90 w_{\parallel}$

$\Rightarrow f_k = 0.90 \times 0.43\,\text{N}$

$\Rightarrow f_k = 0.39\,\text{N}$

$\Sigma F = w_{\parallel} - f_k = ma$

$\Rightarrow w_{\parallel} - 0.90 w_{\parallel} = ma$

$\Rightarrow a = \dfrac{0.10 w_{\parallel}}{m}$

$\Rightarrow a = \dfrac{0.10 \times 0.43\,\text{N}}{4.5 \times 10^{-3}\,\text{kg}}$

$\Rightarrow a = 9.6\,\text{m s}^{-2}$

2 a

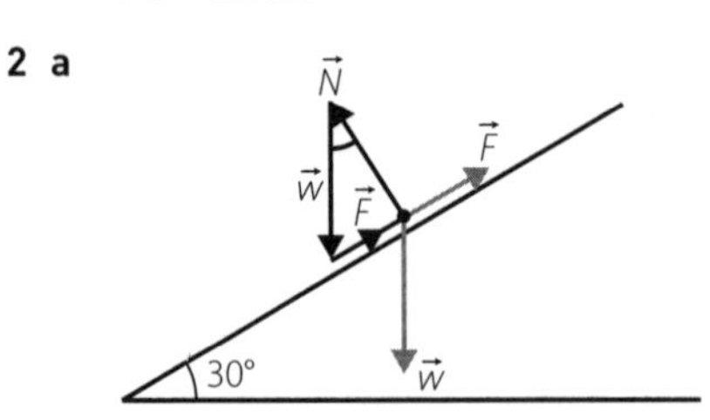

b i $w_{\parallel} = mg\sin\theta$

$\Rightarrow w_{\parallel} = 10\,\text{kg} \times 9.8\,\text{m s}^{-2} \times \sin 30°$

$\Rightarrow w_{\parallel} = 49\,\text{N}$

ii $w_{\perp} = mg\cos\theta$

$\Rightarrow w_{\perp} = 10\,\text{kg} \times 9.8\,\text{m s}^{-2} \times \cos 30°$

$\Rightarrow w_{\perp} = 85\,\text{N}$

9780170412643

c $T - w_{\parallel} = ma$

$\Rightarrow T = w_{\parallel}\ (a = 0\,\text{m s}^{-2})$

$\Rightarrow T = 49\,\text{N}$

3 a *Suvat*:

$s = 56\,\text{m},\ u = 7.2\,\text{km h}^{-1} = \dfrac{7.2\,\text{km h}^{-1}}{3.6\,\text{m}^{-1}\,\text{s}} = 2.0\,\text{m s}^{-1},$

$v = ?,\ a = ?,\ t = 5.6\,\text{s}$

$s = ut + \dfrac{1}{2}at^2$

$\Rightarrow 56\,\text{m} = 2.0\,\text{m s}^{-1} \times 5.6\,\text{s} + \dfrac{1}{2}a \times (5.6\,\text{s})^2$

$\Rightarrow 56\,\text{m} = 11.2\,\text{m} + 18a$

$\Rightarrow 18a = 56\,\text{m} - 11.2\,\text{m} = 44.8\,\text{m}$

$\Rightarrow a = 2.5\,\text{m s}^{-2}$

b $v = u + at$

$\Rightarrow v = 2.0\,\text{m s}^{-1} + 2.5\,\text{m s}^{-2} \times 6.0\,\text{s}$

$\Rightarrow v = 17\,\text{m s}^{-1}$

c Parallel to the slope:

$\Sigma F = w_{\parallel} - f_k = ma$

$\Rightarrow mg\sin\theta - f_k = ma$

$\Rightarrow \sin\theta = \dfrac{ma + f_k}{mg}$

$\Rightarrow \theta = \sin^{-1}\left(\dfrac{ma + f_k}{mg}\right)$

$\Rightarrow \theta = \sin^{-1}\left(\dfrac{1.2\times10^3\,\text{kg}\times 2.5\,\text{m s}^{-2} + 1.0\times10^3\,\text{N}}{1.2\times10^3\,\text{kg}\times 9.8\,\text{m s}^{-2}}\right)$

$\Rightarrow \theta = 20°$

4 a Parallel to surface:

$mg\sin\theta = ma$

$\Rightarrow \theta = \sin^{-1}\left(\dfrac{a}{g}\right)$

$\Rightarrow \theta = \sin^{-1}\left(\dfrac{4.0\,\text{m s}^{-2}}{9.8\,\text{m s}^{-2}}\right)$

$\Rightarrow \theta = 24°$

b Parallel to surface:

$mg\sin\theta - F_{\parallel} = ma$

$\Rightarrow F_{\parallel} = mg\sin\theta - ma$

but, $F_{\parallel} = F\cos\alpha$ (angle relative to slope)

$\Rightarrow F\cos\alpha = mg\sin\theta - ma$

$\Rightarrow F = \dfrac{mg\sin\theta - ma}{\cos\alpha}$

$\Rightarrow F = \dfrac{40\,\text{kg}\times 9.8\,\text{m s}^{-2}\times\sin 30° - 40\,\text{kg}\times 2.5\,\text{m s}^{-2}}{\cos 10°}$

$\Rightarrow F = 97.5\,\text{N}$

5 a i $\Sigma F = mg\sin\theta - f_k = ma$

$\Rightarrow \theta = \sin^{-1}\left(\dfrac{ma + f_k}{mg}\right)$

$\Rightarrow \theta = \sin^{-1}\left(\dfrac{0.45\,\text{kg}\times 0.61\,\text{m s}^{-2} + 0.5\,\text{N}}{0.45\,\text{kg}\times 9.8\,\text{m s}^{-2}}\right)$

$\Rightarrow \theta = 10°$

ii $\Sigma F = mg\sin\theta - f_s = ma$

$\Rightarrow f_s = mg\sin\theta - ma$

$\Rightarrow f_s = 0.45\,\text{kg}\times 9.8\,\text{m s}^{-2}\times\sin 10° - 0.45\,\text{kg}\times 0.61\,\text{m s}^{-2}$

$\Rightarrow f_s = 0.49\,\text{N}$

b $\Sigma F = mg\sin\theta - f_k = ma$

$\Rightarrow a = g\sin\theta - \dfrac{f_k}{m}$

$\Rightarrow a = 9.8\,\text{m s}^{-2}\times\sin 10° - \dfrac{0.5\,\text{N}}{0.45\,\text{kg}}$

$\Rightarrow a = 0.59\,\text{m s}^{-2}$

$\Rightarrow v = u + a\Delta t$

$\Rightarrow \Delta t = \dfrac{v - u}{a} = \dfrac{3.1\,\text{m s}^{-1} - 0\,\text{m s}^{-1}}{0.59\,\text{m s}^{-2}}$

$\Rightarrow \Delta t = 5.3\,\text{s}$

CHAPTER 3 EVALUATION

MULTIPLE-CHOICE

1 B

2 A

3 A

4 C

5 D

6 A

7 D

8 B

9 C

10 D

11 B

12 A

13 A

14 D

15 D

16 A

17 C

18 C

19 B

20 D

SHORT ANSWER

21 a i $N - mg\cos\theta = 0$

$\Rightarrow N = mg\cos\theta$

$\Rightarrow N = 6.0\times10^{-3}\,\text{kg}\times9.8\,\text{m s}^{-2}\times\cos17°$

$\Rightarrow N = 5.6\times10^{-2}\,\text{N}$

ii $w_{\|} = mg\sin\theta$

$\Rightarrow w_{\|} = 6.0\times10^{-3}\,\text{kg}\times9.8\,\text{m s}^{-2}\times\sin17°$

$\Rightarrow w_{\|} = 1.7\times10^{-2}\,\text{N}$

iii $\Sigma F = w_{\|} = mg\sin\theta = ma$

$\Rightarrow a = g\sin\theta$

$\Rightarrow a = 9.8\,\text{m s}^{-2}\times\sin17°$

$a = 2.9\,\text{m s}^{-2}$

b Parallel to plane:

$T_{\|} - w_{\|} = 0$

$\Rightarrow T_{\|} = mg\sin\theta$

but, $T_{\|} = T\sin\theta$

$\Rightarrow T\sin\theta = mg\sin\theta$

$\Rightarrow T = mg\dfrac{\sin\theta}{\cos\theta}$

$\Rightarrow T = mg\times\tan\theta$

$\Rightarrow T = 6.0\times10^{-3}\,\text{kg}\times9.8\,\text{m s}^{-2}\times\tan17°$

$\Rightarrow T = 1.8\times10^{-2}\,\text{N}$

22 a Perpendicular to plane:

$N + 15\sin15° - 10\,\text{kg}\times9.8\,\text{m s}^{-2}\times\cos35° = 0$

$\Rightarrow N = 10\,\text{kg}\times9.8\,\text{m s}^{-2}\times\cos35° - 15\sin15°$

$\Rightarrow N = 76.4\,\text{N}$

b Parallel to plane:

$\Sigma F = T_{\|} - w_{\|} - f$

$\Rightarrow \Sigma F = 15\,\text{N}\times\cos15° - mg\sin\theta° - 5.0\,\text{N}$

$\Rightarrow \Sigma F = 14.5\,\text{N} - 6.0\times10^{-3}\,\text{kg}\times9.8\,\text{m s}^{-2}\times\sin15° - 5.0\,\text{N}$

$\Sigma F = 9.5\,\text{N}$

c $\Sigma F = ma$

$\Rightarrow \Sigma F = 9.5\,\text{N}$

$\Rightarrow a = \dfrac{9.5\,\text{N}}{6.0\times10^{-3}\,\text{kg}}$

$\Rightarrow a = 1.6\times10^{3}\,\text{m s}^{-2}$

23 a i $N = mg\cos\theta$

$\Rightarrow N = 2.5\,\text{kg}\times9.8\,\text{m s}^{-2}\times\cos22°$

$\Rightarrow N = 22.7\,\text{N}$

ii Parallel to plane:

$f_s = mg\sin\theta$

$\Rightarrow f_s = 2.5\,\text{kg}\times9.8\,\text{m s}^{-2}\times\sin22°$

$\Rightarrow f_s = 9.2\,\text{N}$

iii $f_k = 0.40 * \text{N}$

$\Rightarrow f_k = 0.40 * 22.7\,\text{N}$

$\Rightarrow f_k = 9.1\,\text{N}$

b Parallel to plane:

$w_{\|} - f_k = ma$

$\Rightarrow mg\sin\theta - f_k = ma$

$\Rightarrow a = mg\sin\theta - 0.40\times mg\cos\theta$

$\Rightarrow a = g\sin\theta - 0.40\times g\cos\theta$

$\Rightarrow a = 9.8\,\text{m s}^{-2}\times(\sin30° - 0.40\times g\cos\theta)$

$\Rightarrow a = 9.8\,\text{m s}^{-2}\times(\sin30° - 0.40\times\cos\theta)$

$\Rightarrow a = 1.5\,\text{m s}^{-2}$

$s = 8.0\,\text{m};\ u = 0\,\text{m s}^{-1};\ a = 1.5\,\text{m s}^{-2};\ t = ?$

$s = ut + \dfrac{1}{2}at^2$

$\Rightarrow t = \sqrt{\dfrac{2s}{a}}$

$\Rightarrow t = \sqrt{\dfrac{2\times8.0\,\text{m}}{1.5\,\text{m s}^{-2}}}$

$\Rightarrow t = 3.3\,\text{s}$

24 a $\Sigma F_{\|,\text{down}} = mg\sin\theta - f_k$

$\Rightarrow \Sigma F_{\|,\text{down}} = 5.0\,\text{kg}\times9.8\,\text{m s}^{-2}\times\sin24° - 15\,\text{N}$

$\Rightarrow \Sigma F_{\|,\text{down}} = 4.9\,\text{N}$

b System analysis:

$w_{\text{vertical}} - (w_{\|} - f_k) = (M+m)a$

$\Rightarrow 2.5\,\text{kg}\times9.8\,\text{m s}^{-2} - 4.9\,\text{N} = 10a$

$\Rightarrow a = \dfrac{49\,\text{N} - 4.9\,\text{N}}{10}$

$\Rightarrow a = 4.4\,\text{m s}^{-2}$

OR

Parallel to slope:

$T - 4.9\,N = 5.0\,kg\times a$

$\Rightarrow T - 4.9 = 5a \quad (1)$

Vertically:

$mg - T = 5.0\,kg\times a$

$\Rightarrow 5.0\,\text{kg}\times9.8\,\text{m s}^{-2} - T = 5a$

$\Rightarrow 49 - T = 5a \quad (2)$

(1) + (2):

$\Rightarrow 10a = 44.1$

$\Rightarrow a = 4.4\,\text{m s}^{-2}$

c $w_{\text{vertical}} - (w_{\|} + f_k) = (M+m)a$

$w_{\text{vertical}} - \Sigma F_{\|,\text{down}} = (M+m)a$

$\Rightarrow w_{\text{vertical}} - (w_{\|} + f_k) = (M+m)a$

if $a = 0$

$\Rightarrow w_{\text{vertical}} = w_{\|} + f_k$

$\Rightarrow 49\,\text{N} = 49\sin\theta + 15\,\text{N}$

$\Rightarrow \sin\theta = \dfrac{34}{49}$

$\Rightarrow \theta = 44°$

CHAPTER 4 REVISION

4.1 UNIFORM CIRCULAR MOTION

1 Both period and frequency refer to the same periodic motion. Period is the time taken for one revolution; frequency is the number of times the motion repeats in a given time interval, usually one second. They are the inverse of each other.

2

r (m)	T (s)	f (s^{-1})	v (m s^{-1})
1.5	2.0	0.5	4.7
12.7	20	0.05	4.0
0.60	0.25	4.0	15.0

3 a $f = \frac{1}{T}$

$f = \frac{5.0}{44\,\text{s}}$

$f = 0.11\,\text{Hz}$

b $v = 2\pi rf$

$v = 2\pi \times 1.8\,\text{m} \times 0.11\,\text{Hz}$

$v = 1.3\,\text{m s}^{-1}$

4 a $T = \frac{1.0\,\text{min} \times 60\,\text{s min}^{-1}}{12\,\text{rev}}$

$T = 5\,\text{s}$

b $v = \frac{2\pi r}{T}$

$v = \frac{2\pi \times 6.20\,\text{m}}{5.0\,\text{s}}$

$v = 7.8\,\text{m s}^{-1}$

5 $v = \frac{2\pi r}{T}$

$v = \frac{2\pi\left(\frac{d}{2}\right)}{T}$

$v = \frac{\pi \times d}{T}$

$d = \frac{vT}{\pi}$

$d = \frac{\left(240\,\text{km h}^{-1} \times 10^3\,\text{m km}^{-1} \times \frac{1}{3600\,\text{h s}^{-1}}\right) \times 52.0\,\text{s}}{\pi}$

$d = 1.1\,\text{km}$

4.2 CENTRIPETAL ACCELERATION AND FORCE

1

r (m)	v (m s^{-1})	a (m s^{-2})
1.5	4.0	11
3.5	5.8	9.6
16	15.0	14

2

r (m)	T (s)	f (s^{-1})	a (m s^{-2})
63	40	(–)	1.6
3.5	(–)	0.20	5.5
76	19	(–)	8.3
1.2×10^{-3}	(–)	3.1	0.47

3 a $a = \frac{4\pi^2 r}{T^2}$

$a = \frac{4\pi^2 \times 3.84 \times 10^8\,\text{m}}{\left(27.3\,\text{d} \times 24\,\text{h d}^{-1} \times 3600\,\text{s h}^{-1}\right)^2}$

$a = 2.7 \times 10^{-3}\,\text{m s}^{-2}$

b $a = \frac{2\pi v}{T}$

$v = \frac{aT}{2\pi}$

$v = \frac{2.7 \times 10^{-3}\,\text{m s}^{-2} \times 27.3\,\text{d} \times 24\,\text{h d}^{-1} \times 3600\,\text{s h}^{-1}}{2\pi}$

$v = 1.0 \times 10^3\,\text{m s}^{-1}$

4 a $a = \frac{v^2}{r}$

$a = \frac{\left(2.41 \times 10^3\,\text{m s}^{-1}\right)^2}{2.28 \times 10^{11}\,\text{m}}$

$a = 2.55 \times 10^{-5}\,\text{m s}^{-2}$

b $a = \frac{4\pi^2 r}{T^2}$

$T^2 = \frac{4\pi^2 r}{a}$

$T = 2\pi\sqrt{\frac{r}{a}}$

$T = 2\pi\sqrt{\frac{2.28 \times 10^{11}\,\text{m}}{2.55 \times 10^{-5}\,\text{m s}^{-2}}}$

$T = 8.94 \times 10^{15}\,\text{s}$

5 $T = \frac{24\,\text{h} \times 3600\,\text{s h}^{-1}}{6.32\,\text{rev}}$

$T = 1.37 \times 10^4\,\text{s}$

$a = \frac{2\pi v}{T}$

$v = \frac{aT}{2\pi}$

$v = \frac{2.6\,\text{m s}^{-2} \times 1.37 \times 10^4\,\text{s}}{2\pi}$

$v = 5.66 \times 10^3\,\text{m s}^{-1}$

4.3 NET FORCE CAUSES CIRCULAR MOTION

1 a F(by wall of rotor on person)

b $\Sigma F = m \times 4\pi^2 r f^2$

$\Sigma F = 60\,\text{kg} \times 4\pi^2 \times 7.0\,\text{m} \times (0.20\,\text{Hz})^2$

$\Sigma F = 663\,\text{N}$

2 $\Sigma F = m\frac{4\pi^2 r}{T^2}$

$T^2 = m\frac{4\pi^2 r}{\Sigma F}$

$T = 2\pi\sqrt{\frac{mr}{\Sigma F}}$

$T = 2\pi\sqrt{\frac{0.150\,\text{kg} \times 0.35\,\text{m}}{67\,\text{N}}}$

$T = 0.18\,\text{s}$

3 $\Sigma F = m\frac{v^2}{r}$

$\Sigma F = 450\,\text{kg} \times \frac{\left(95\,\text{km}\,\text{h}^{-1} \times 10^3\,\text{m}\,\text{km}^{-1} \times \frac{1}{3600}\,\text{h}\,\text{s}^{-1}\right)^2}{65\,\text{m}}$

$\Sigma F = 4.8\,\text{kN}$

4 $\Sigma F = m \times 4\pi^2 r f^2$

$f^2 = \frac{\Sigma F}{m \times 4\pi^2 r}$

$f = \frac{1}{2\pi}\sqrt{\frac{\Sigma F}{mr}}$

$f = \frac{1}{2\pi}\sqrt{\frac{8.2 \times 10^{-8}\,\text{N}}{9.1 \times 10^{-31}\,\text{kg} \times 5.3 \times 10^{-9}\,\text{m}}}$

$f = 6.6 \times 10^{14}\,\text{Hz}$

5

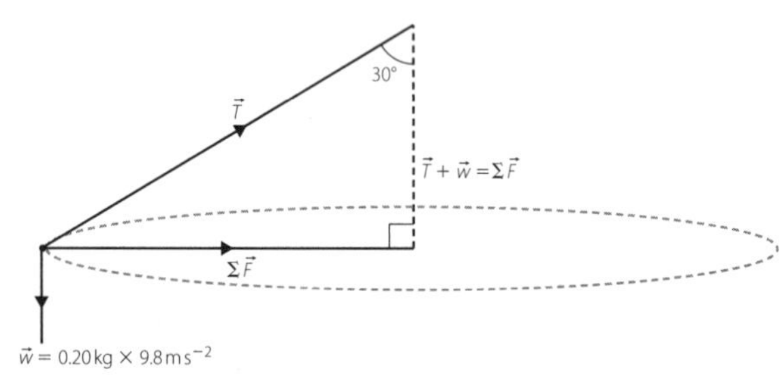

$\Sigma F = m\frac{4\pi^2 r}{T^2}$

$T = \frac{20\,\text{s}}{23.2\,\text{rev}} = 0.862\,\text{s}$

$r = R\sin\theta = 0.80\,\text{m} \times \sin 30° = 0.40\,\text{m}$

$\Sigma F = 0.200\,\text{kg} \times \frac{4\pi^2 \times 0.40\,\text{m}}{(0.862\,\text{s})^2}$

$\Sigma F = 4.3\,\text{N}$

4.4 UNIFORM MOTION ON NON-HORIZONTAL SURFACES

1

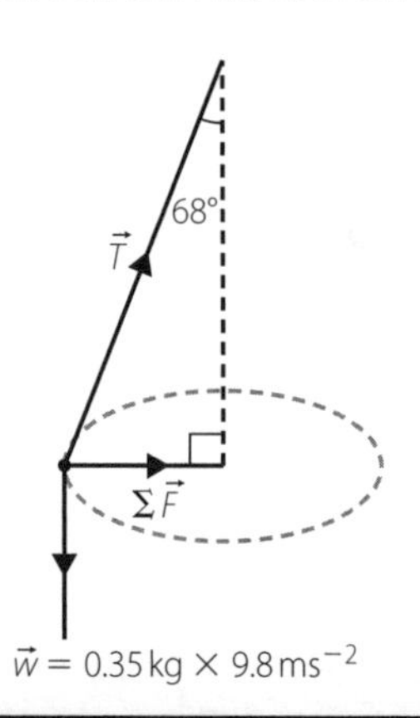

Horizontally:

$\Sigma F = T\sin\theta = T\sin 68°$

Vertically:

$T\cos\theta - mg = 0$

$T = \frac{mg}{\cos\theta}$

Combined:

$\Sigma F = \frac{mg}{\cos\theta}\sin\theta$

$\Sigma F = mg\tan\theta$

$\Sigma F = 0.35\,\text{kg} \times 9.8\,\text{m}\,\text{s}^{-2} \times \tan 68°$

$\Sigma F = 8.5\,\text{N}$

2

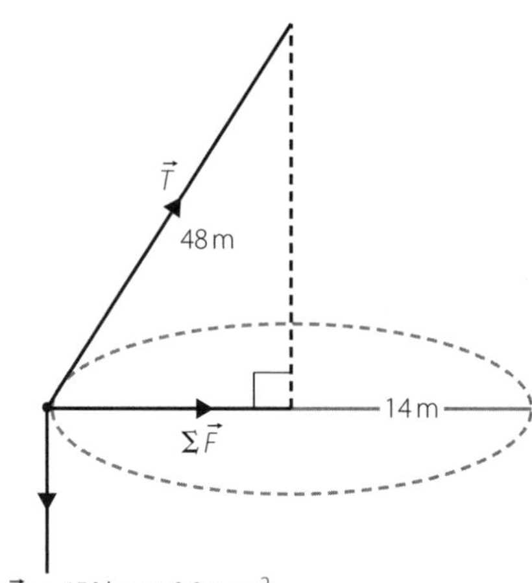

$\Sigma F = mg\tan\theta$

$\tan\theta = \frac{14\,\text{m}}{\sqrt{(48\,\text{m})^2 - (14\,\text{m})^2}} = 0.31$

$\Sigma F = 450\,\text{kg} \times 9.8\,\text{m}\,\text{s}^{-2} \times 0.31$

$\Sigma F = 1.3 \times 10^3\,\text{N}$

3

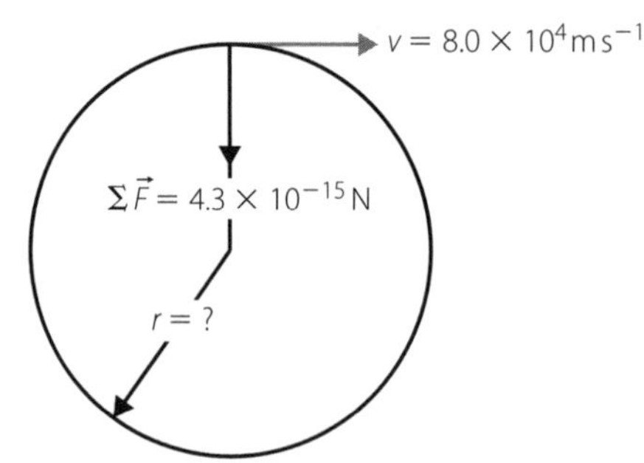

$\Sigma F = m\frac{v^2}{r}$

$r = m\frac{v^2}{\Sigma F}$

$r = 4.1 \times 10^{-26}\,\text{kg} \times \frac{(8.0 \times 10^4\,\text{m}\,\text{s}^{-1})^2}{4.3 \times 10^{-15}\,\text{N}}$

$r = 6.1\,\text{cm}$

9780170412643

4 $T = \dfrac{16\,\text{s}}{20\text{ turns}} = 0.80\,\text{s}$

$$\Sigma F = m \times \frac{4\pi^2 r}{T^2}$$

$$\Sigma F = 45\,\text{kg} \times \frac{4\pi^2 \times 5.6\,\text{m}}{(0.80\,\text{s})^2}$$

$$\Sigma F = 1.6 \times 10^4\,\text{N}$$

5

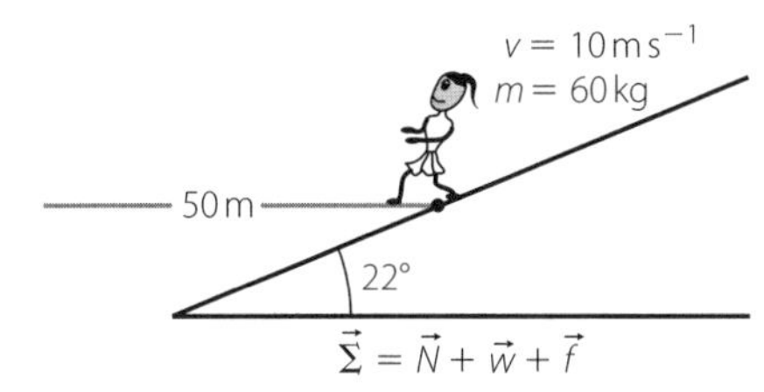

a $\Sigma F = m\dfrac{v^2}{r}$

$$\Sigma F = 60.0\,\text{kg} \times \frac{(10.0\,\text{m s}^{-1})^2}{50\,\text{m}}$$

$$\Sigma F = 120\,\text{N}$$

b Without friction:

$$\Sigma F = mg\tan\theta$$

$$\Sigma F = 60\,\text{kg} \times 9.8\,\text{m s}^{-2} \times \tan 22°$$

$$\Sigma F = 238\,\text{N}$$

Friction up the slope is required to maintain this speed.

6

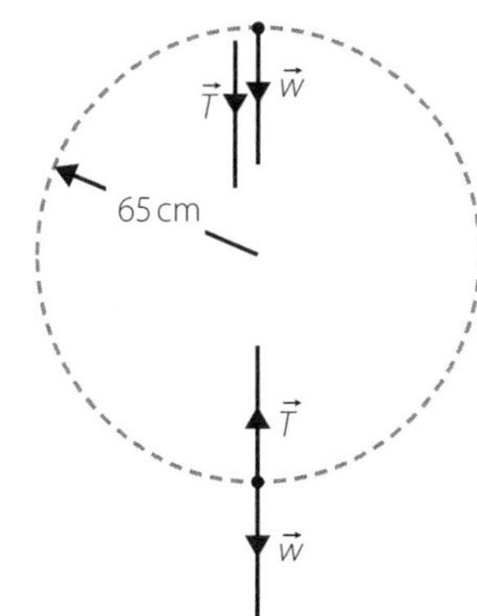

a $T + w = m\dfrac{v^2}{r}$

$$T = m\left(\frac{v^2}{r} - g\right)$$

$$T = 80\,\text{kg} \times \left(\frac{(7.3\,\text{m s}^{-1})^2}{0.65\,\text{m}} - 9.8\,\text{m s}^{-2}\right)$$

$$T = 5.8\,\text{kN}$$

b $T - w = m\dfrac{v^2}{r}$

$$T - m\left(\frac{v^2}{r} + g\right)$$

$$T = 80\,\text{kg} \times \left(\frac{(7.3\,\text{m s}^{-1})^2}{0.65\,\text{m}} + 9.8\,\text{m s}^{-2}\right)$$

$$T = 7.3\,\text{kN}$$

7 $v_{max} = \sqrt{gr}$

$$v_{max} = \sqrt{9.8\,\text{m s}^{-2} \times 2.1\,\text{m}}$$

$$v_{max} = 4.5\,\text{m s}^{-1} \times 10^{-3}\,\text{km m}^{-1} \times 3600\,\text{s h}^{-1}$$

$$v_{max} = 16.3\,\text{km h}^{-1}$$

Alternatively:

$$\Sigma F = mg - N = m\frac{v^2}{r}$$

$$N = m\left(g - \frac{v^2}{r}\right)$$

$$N = 105\,\text{kg} \times \left(9.8\,\text{m s}^{-2} - \frac{\left(25\,\text{km h}^{-1} \times 10^3\,\text{m km}^{-1} \times \frac{1}{3600}\,\text{h s}^{-1}\right)^2}{2.1\,\text{m}}\right)$$

$$N = 105\,\text{kg} \times \text{m s}^{-2} \times {}^{-}13.2\,\text{m s}^{-2}$$

$N < 0$; cyclist leaves ground

8 $\Sigma F = N - mg = m\dfrac{v^2}{r}$

$$N = mg + m\frac{v^2}{r}$$

$$N = 3g$$

$$3mg = mg + m\frac{v^2}{r}$$

$$\frac{v^2}{r} = 2g$$

$$v = \sqrt{2gr}$$

$$v = \sqrt{2 \times 9.8\,\text{m s}^{-2} \times 25\,\text{m}}$$

$$v = 22\,\text{m s}^{-1}$$

CHAPTER 4 EVALUATION

MULTIPLE-CHOICE

1 A
2 A
3 D
4 B
5 A
6 A
7 A
8 D
9 C
10 B
11 D
12 B
13 A
14 C
15 B
16 A
17 D

18 A

19 A

20 C

SHORT ANSWER

21 a

b Horizontally: $T\sin\theta = \Sigma F$

Vertically: $T\cos\theta = mg$

Combined: $\frac{T\sin\theta}{T\cos\theta} = \frac{\Sigma F}{mg}$

$\Sigma F = mg\tan\theta$

$\tan\theta = \frac{16\,\text{cm}}{\sqrt{(80\,\text{cm})^2 - (16\,\text{cm})^2}}$

$\tan\theta = 0.20$

$\Sigma F = 0.20\,\text{kg} \times 9.8\,\text{m}\,\text{s}^{-2} \times 0.20$

$\Sigma F = 0.40\,\text{N}$

c Vertically: $T\cos\theta = mg$

$\cos\theta = \frac{\sqrt{(80\,\text{cm})^2 - (16\,\text{cm})^2}}{80\,\text{cm}}$

$\cos\theta = 0.98$

$T = \frac{mg}{\cos\theta}$

$T = \frac{0.20\,\text{kg} \times 9.8\,\text{m}\,\text{s}^{-2}}{0.98}$

$T = 2.0\,\text{N}$

d $\Sigma F = m4\pi^2 rf^2$

$f = \frac{1}{2\pi}\sqrt{\frac{\Sigma F}{mr}}$

$f = \frac{1}{2\pi}\sqrt{\frac{0.40\,\text{N}}{0.20\,\text{kg} \times 0.16\,\text{m}}}$

$f = 0.56\,\text{Hz}$

22 a

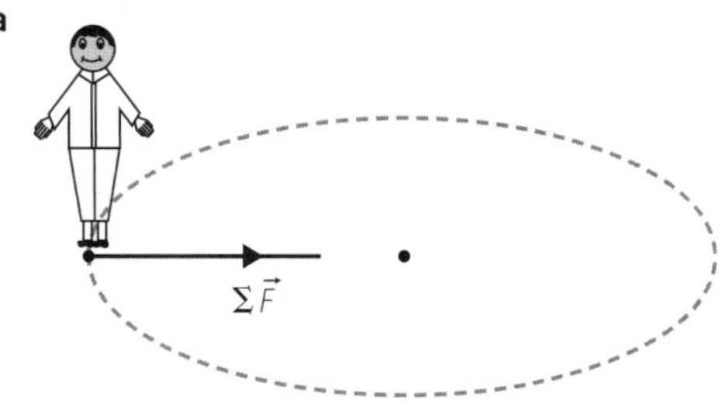

b $v = \frac{2\pi r}{T}$

$T = \frac{2\pi r}{v}$

$T = \frac{2\pi \times 4.5\,\text{m}}{3.0\,\text{m}\,\text{s}^{-1}}$

$t(30°) = \frac{30°}{360°} \times T$

$t(30°) = \frac{30°}{360°} \times \frac{2\pi \times 4.5\,\text{m}}{3.0\,\text{m}\,\text{s}^{-1}}$

$t(30°) = 0.79\,\text{s}$

c $\Sigma F = m\frac{v^2}{r}$

$\Sigma F = 60\,\text{kg} \times \frac{(3.0\,\text{m}\,\text{s}^{-1})^2}{4.5\,\text{m}}$

$\Sigma F = 120\,\text{N}$

d $\frac{\Sigma F}{mg} = \tan\theta$

$\theta = \tan^{-1}\left(\frac{\Sigma F}{mg}\right)$

$\theta = \tan^{-1}\left(\frac{120\,\text{N}}{60\,\text{kg} \times 9.8\,\text{m}\,\text{s}^{-2}}\right)$

$\theta = 11.5°$

23 a $\Sigma F = N - w = 5mg - mg = 4mg$

$m\frac{v^2}{r} = 4mg$

$v = 2\sqrt{gr}$

$v = 2\sqrt{9.8\,\text{m}\,\text{s}^{-2} \times 0.50\,\text{m}}$

$v = 4.4\,\text{m}\,\text{s}^{-1}$

b $\Sigma F = N + w = m\frac{v^2}{r}$

$N + mg = m\frac{v^2}{r}$

When $N = 0$:

$v = \sqrt{gr}$

$v = \sqrt{9.8\,\text{m}\,\text{s}^{-2} \times 0.50\,\text{m}}$

$v = 2.2\,\text{m}\,\text{s}^{-1}$

c $\Sigma\vec{F} = \vec{N} + \vec{w}$

$\vec{N}$ is always directed along the radius

$\Rightarrow \vec{N} + \vec{w}$ cannot be centre-seeking, except when aligned at top and bottom.

24 a $\frac{w}{N} = \cos\theta$

$N = \frac{mg}{\cos\theta}$

$N = \frac{1000\,\text{kg} \times 9.8\,\text{m}\,\text{s}^{-2}}{\cos 18°}$

$N = 1.0 \times 10^4\,\text{N}$

b $\Sigma F = mg\tan\theta$

$\Sigma F = 1000\,\text{kg} \times 9.8\,\text{m}\,\text{s}^{-2} \times \tan 18°$

$\Sigma F = 3.0 \times 10^4\,\text{N}$

c $\Sigma F = m\dfrac{v^2}{r}$

$r = \dfrac{mv^2}{\Sigma F}$

$r = \dfrac{1.0 \times 10^3\,\text{kg} \times \left(8.0\,\text{m}\,\text{s}^{-1}\right)^2}{1.0 \times 10^4\,\text{N}}$

$r = 6.4\,\text{m}$

25 a $\Sigma F = ma, a = \dfrac{\Sigma F}{m} = \dfrac{1.0 \times 10^4\,\text{N}}{1.0 \times 10^3\,\text{kg}}, a = 10\,\text{m}\,\text{s}^{-2}$

$a \propto \dfrac{1}{r}$

when $r \rightarrow \dfrac{1}{2}r, a \rightarrow 2a : a = 20\,\text{m}\,\text{s}^{-2}$

b No change to speed:

$a = \dfrac{v^2}{r}$

$v = \sqrt{ar}$

if $a \times 2$ and $r \times \dfrac{r}{2}, v = \sqrt{2a \times \dfrac{r}{2}} = \sqrt{ar} = 8.0\,\text{m}\,\text{s}^{-1}$

CHAPTER 5 REVISION

5.1 GRAVITATIONAL FORCE AND FIELD KEY TERMS

1 Newton's law of universal gravitation relates the force of gravity acting between two objects with their individual masses and the distance between them. The gravitational potential energy possessed by an object in a gravitational field may be determined using GPE = *mgn*, which incorporates Newton's second law.

2 Student's answers will vary

5.2 THE HISTORY OF GRAVITY

1 a Aristotle recognised that objects feel due to their inherent heaviness, termed gravitas.

b Newton developed the description of gravitational force to involve a relationship between force, mass and distance that obeyed an inverse-square law. Newton's understanding of gravity was built on the impressive measurements of Brahe and the mathematical interpretation of these data by Kepler.

c Galileo conducted experiments on projectiles and falling objects showing that falling objects accelerated uniformly.

d Einstein made significant modifications to our definition of gravitation in his 1915 paper on general relativity, most notably, the consideration of four-dimensional space-time.

2 $F = G\dfrac{m_1 m_2}{r^2}$. Author to supply full answer.

3 a Decreases

b Increases

c Remains the same

4 a Work

b Potential energy

c Gravitas

d Galileo

e Empirical

f Kinematics

5.3 GRAVITATIONAL POTENTIAL ENERGY

1 14112 J

2 a 2.5 m

b $3.063 \times 10^3\,\text{J}$

3 a 1323 J

b $5.94\,\text{m}\,\text{s}^{-1}$

4 a $3.51 \times 10^9\,\text{J}$

b $9.03 \times 10^{10}\,\text{J}$

c $9.38 \times 10^{10}\,\text{J}$

5 a 470400 J

b $24.2\,\text{m}\,\text{s}^{-1}$

5.4 GRAVITATIONAL FIELDS

1 1470 N – 553.5 N = 916.5 N

2 8089.9 N

3 38.7 N

4 $6.00 \times 10^{-8}\,\text{N}$

5.5 GRAVITATIONAL FIELD STRENGTH

1 15.6 m

2 $1.32\,\text{m}\,\text{s}^{-2}$

3 a 11.48 m

b $1.53 \times 2 = 3.06\,\text{s}$

4 $g_{\text{Europa}} = 1.32\,\text{m}\,\text{s}^{-2}$

$g_{\text{Io}} = 1.78\,\text{m}\,\text{s}^{-2}$

$g_{\text{Callisto}} = 1.24\,\text{m}\,\text{s}^{-2}$

$g_{\text{Ganymede}} = 1.43\,\text{m}\,\text{s}^{-2}$

5.6 NEWTON'S LAW OF UNIVERSAL GRAVITATION AND GRAVITATIONAL FORCE

1 $g = \dfrac{F_G}{m}; F_G = \dfrac{G \times m_1 \times m_2}{r^2}$

The field exists independent of mass. The field is measured in $\text{m}\,\text{s}^{-2}$ or $\text{N}\,\text{kg}^{-1}$; whereas, the force is measured in Newtons.

2 1.50 F

3 $3.43 \times 10^{21}\,\text{N}$

4 $F_{\text{G Earth}} = 159.28\,\text{N}$

$F_{\text{G Moon}} - 0.24\,\text{N}$

$F_{\text{G net}} = 159.04\,\text{N}$ toward Earth

CHAPTER 5 EVALUATION

MULTIPLE-CHOICE

1 D

2 D

SHORT ANSWER

3 24612 N

4 $F_{G\ Moon-Earth} = 1.98 \times 10^{20}$ N

$F_{G\ Sun-Earth} = 3.52 \times 10^{22}$ N

5 Gravitational field describes the field in space that exists surrounding objects with mass. The field is measured in $N\,kg^{-1}$ or $m\,s^{-2}$. Gravitational force describes the force that acts at a distance between all objects with mass. The force is measured in Newtons.

6 $F_G = 2.67 \times 10^{-10}$ N. At twice the distance the force becomes one quarter the strength, to $F_G = 6.67 \times 10^{-11}$ N.

7 $T_f = 2 \times 0.816\,s = 1.63\,s$

8 **a** $g = 3.31\,N\,kg^{-1}$

b $3.31\,m\,s^{-2}$

c 165.5 J

d 165.5 J

9 1764 J

10 **a** 1.42 s

b The period decreases

11 $44.27\,m\,s^{-1}$

CHAPTER 6 REVISION

6.1 ORBITAL MOTION KEY TERMS

1 Johannes Kepler determined three laws to describe the motion of planets: All planets move in elliptical orbits with the Sun at one focus. The second law reflects that a line connecting the Sun and a planet sweeps out equal areas in equal periods of time. Kepler's third law relates the period of orbit and the mean orbital distance. Should an object travel at a critical orbit velocity then it will remain in orbit at the same altitude and experience weightlessness.

2 Student's answers will vary.

6.2 KEPLER'S LAWS OF PLANETARY MOTION

1 Kepler's first law (the law of ellipses): all planets move in elliptical orbits with the Sun at one focus.

Kepler's second law (the law of equal areas): a line that connects a planet to the Sun sweeps out equal areas in equal time periods.

Kepler's third law (the law of periods): the square of the period of a planet's orbit is proportional to the cube of its mean orbital distance.

2 $Radius_{Europa} = 6.68 \times 10^8$ m; $Period_{Ganymede} = 7.11$ days; $Radius_{Callisto} = 1.87 \times 10^9$ m

3 3.75×10^8 s.

4 Average $\frac{R^3}{T^2} = 1.033 \times 10^{19}$

Planet C $\frac{R^3}{T^2} = 1.086 \times 10^{19}$

Planet C may be confirmed as part of this system.

6.3 NEWTON'S LAW OF UNIVERSAL GRAVITATION AND KEPLER'S THIRD LAW.

1 $F_C = F_G$

$$\frac{mv^2}{r} = \frac{Gmm}{r^2}$$

$$v^2 = \frac{Gm}{r}$$

$$v = \sqrt{\frac{Gm}{r}}$$

2 **a** 1.979×10^{20} km

b 6.405 Mpc

c 6.405×10^6 pc

3 T = 5062 s or 1.406 hours.

4 $T = 7.54 \times 10^6$ s or 2095 hours.

6.4 SATELLITE MOTION

1 3.52×10^{16} N

2 $6.197\,m\,s^{-1}$

3 $v_{Orbit} = 4655.8\,m\,s^{-1}$; $t = 24791$ s or 6.89 hours

4 $7613.3\,m\,s^{-1}$

CHAPTER 6 EVALUATION

MULTIPLE-CHOICE

1 D

2 A

SHORT ANSWER

3 5.20 Earth years

4 6.50×10^{28} kg

5 $1.00 \times 10^{13}\,m^3\,s^{-2}$

6 1 Mpc = 3.09×10^{22} m, 1 ly = 9.47×10^{15} m, 1 AU = 1.50×10^{11} m

Milky Way diameter = 100 000 ly (or 0.03 Mpc); the light year is most appropriate.

7 $t = 33.4$ hours

8 2436.2 N

9 $4126.2\,m\,s^{-1}$

10 2.50×10^7 m

11 Orbital velocity is a precise velocity to keep an object in orbit at a given radius or altitude. Orbital acceleration is the result of a force applied to keep an object in orbit – typically through centripetal force. Escape velocity is the minimum speed required to escape the gravitational influence of a mass.

CHAPTER 7 REVISION

7.1 COULOMB'S LAW

1 $F = \frac{kQq}{r^2}$. The force a charge q experiences from charge Q is proportional to the product of the charges, and inversely proportional to the distance between the two charges squared.

2 Attractive forces occur between two unlike charges, positive and negative. This causes the charges to move towards each other. Repulsive forces occur between two like charges, such as two negative charges or two positive charges. Unlike attractive forces, when a repulsive force is experienced between two charges, they will move away from each other.

3 Electrostatics is the study of stationary charges. Electricity has to do with moving charges (electrons).

4 Electrons will always distribute equally when two charged objects come into contact with each other, assuming electrons are free to move. When move away from each other, the resulting charge on each object will be the same.

5 When taking off a nylon shirt and hair sticks up, photocopying or printers. There are many more examples!

6 **a** ←——• $F_{\text{A on B}}$ •——→ $F_{\text{B on A}}$

b Repulsive

c Yes, as it is experiencing a force from A.

d The force that B experiences will double.

e $F_1 = \frac{kQq}{r_1^2}$

$F_2 = \frac{kQq}{r_2^2}$

$r_2 = \frac{2}{3}r_1$

$F_2 = \frac{kQq}{\left(\frac{2}{3}r_1\right)^2}$

$F_2 = \frac{9}{4}\frac{kQq}{r_1^2}$

$F_2 = \frac{9}{4}F_1$

7 $F = \frac{kQq}{r^2}$

$F = \frac{9\times10^9(1.6\times10^{-19})^2}{(10\times10^{-6})^2}$

$F = 2.30\times10^{-18}\text{ N}$

Repulsive

8 $F = \frac{kQq}{r^2}$

$62 = \frac{9\times10^9\times q^2}{\left(6.5\times10^{-6}\right)^2}$

$q = \sqrt{\frac{62\times\left(6.5\times10^{-6}\right)^2}{9\times10^9}}$

$q = 5.39\times10^{-10}\text{ C}$

9 $F = \frac{kQq}{r^2}$

$33 = \frac{9\times10^9\times5.2\times10^{-6}\times6.8\times10^{-6}}{r^2}$

$r = \sqrt{\frac{9\times10^9\times5.2\times10^{-6}\times6.8\times10^{-6}}{33}}$

$r = 0.0982\text{ m}$

$F = \frac{kQq}{r^2}$

$F = \frac{9\times10^9\times5\times10^{-6}\times10\times10^{-6}}{0.05^2}$

$F = 180\text{ N}$

$E = \frac{F}{q}$

$E = \frac{mg}{q}$

$E = \frac{3.6\times10^{-6}\text{ kg}\times9.81\text{ ms}^{-2}}{2.0\times10^{-6}\text{ C}}$

$E = 17.66\text{ NC}^{-1}$

10

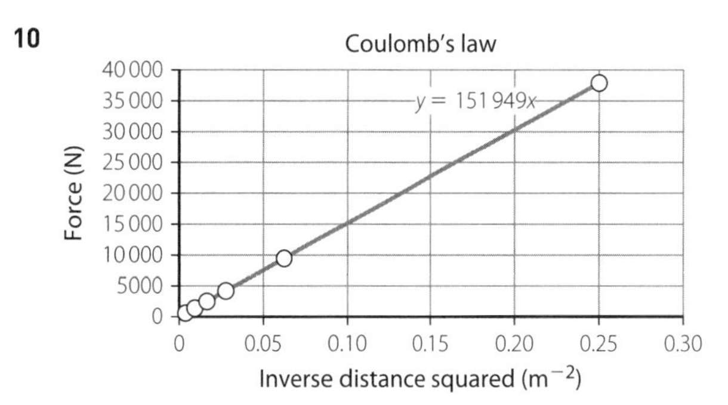

From the gradient:

$m = kqQ$

$k = \frac{m}{qQ}$

$k = \frac{151949}{4.1\times10^{-3}\times4.1\times10^{-3}}$

$k = 9.04\times10^9\text{ Nm}^2\text{ C}^{-2}$

$\therefore \frac{1}{4\pi\varepsilon_0} = 9.04\times10^9\text{ Nm}^2\text{ C}^2$

7.2 ELECTRIC FIELDS

1

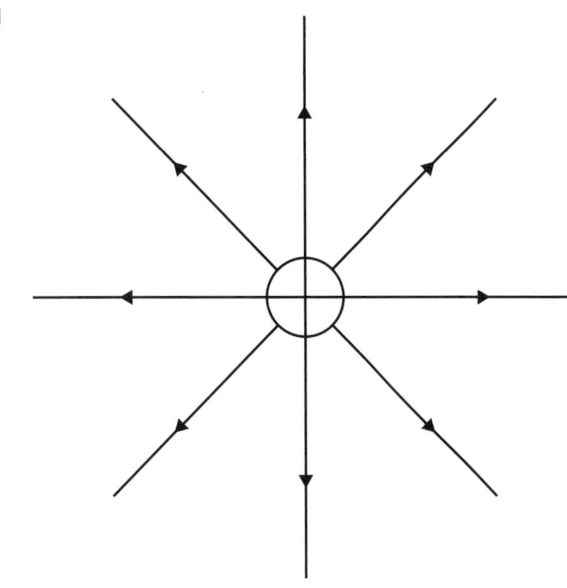

2

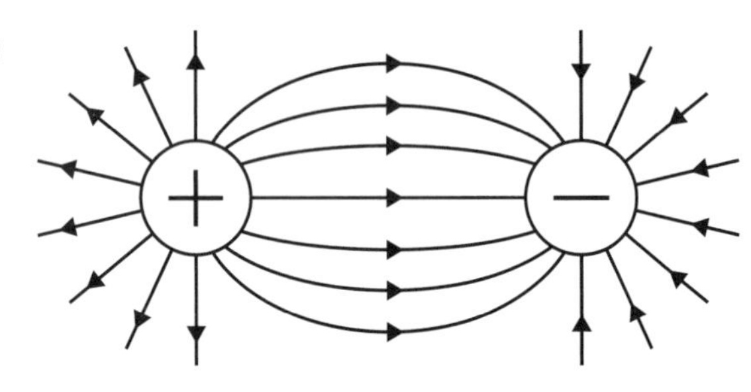

3

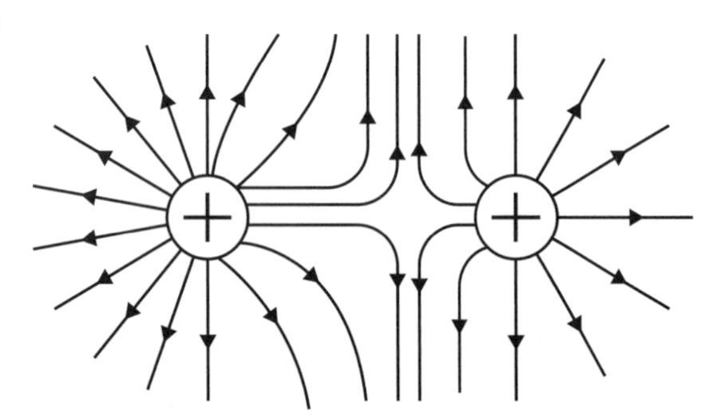

4 It will follow the field lines in the opposite direction to which the field line is pointing. As an electron is a negatively charged particle, it will go in the opposite direction to a positively charged particle (field line direction) but will still follow the same pathway.

5 $E = \frac{F}{q}$

$E = \frac{3.1 \times 10^{-2}\,\text{N}}{4 \times 10^{-6}\,\text{C}}$

$E = 7750\,\text{N}\,\text{C}^{-1}$

6 $E = \frac{1}{4\pi\varepsilon_0}\frac{Q}{r^2}$

$E = 9 \times 10^9\,\text{C}\,\text{m}^2\,\text{C}^{-2} \times \frac{8 \times 10^{-6}\,\text{C}}{(0.1\,\text{m})^2}$

$E = 7.2 \times 10^6\,\text{N}\,\text{C}^{-1}$

7 $E = \frac{1}{4\pi\varepsilon_0}\frac{Q}{r^2}$

$E_{1\text{cm}} = 9 \times 10^9 \times \frac{6 \times 10^{-6}}{0.01^2}$

$E_{1\text{cm}} = 5.4 \times 10^8\,\text{N}\,\text{C}^{-1}$

$E_{10\text{cm}} = 5.4 \times 10^6\,\text{N}\,\text{C}^{-1}$

$E_{1\text{m}} = 5.4 \times 10^4\,\text{N}\,\text{C}^{-1}$

As the distance from the source increases 10-fold, the electric field decreases by two orders of magnitude.

7.3 ELECTRIC POTENTIAL ENERGY

1 Electric potential energy is the energy a charged particle has in an electric field equal to the amount of work it is capable of doing in that electric field.

2 Yes. If electric potential is negative it means that work must be done on that charge in order to move it. For example, if you want a positive charge to move *against* the field lines, work must be performed on it.

3 $\Delta V = q\Delta U$

$\Delta V = 0.7\,\text{C} \times 6.7\,\text{J}$

$\Delta V = 4.69\,\text{V}$

It has a larger potential at point A because there is work being done on the charge to move it from point A to point B.

4 Potential difference between the plates:

$V = 1000\,\text{V}\,\text{m}^{-1} \times 0.1\,\text{m}$

$V = 100\,\text{V}$

If an electron were to accelerate across the potential, then it would obtain $E = qV$ of energy:

$E = 1.6 \times 10^{-19}\,\text{C} \times 100\,\text{V}$

$E = 1.6 \times 10^{-17}\,\text{J}$

This energy is all kinetic energy, and the velocity of the electron can be found as follows:

$E = \frac{1}{2}mv^2$

$v = \sqrt{\frac{2E}{m}}$

$v = \sqrt{\frac{2 \times 1.6 \times 10^{-17}}{9.11 \times 10^{-31}}}$

$v = 5.93 \times 10^5\,\text{m}\,\text{s}^{-1}$

5 Halfway between the plates the charge would have 50V of potential. If the charge was then free to move it would move with the field lines if it was a positive charge or against the field lines if it was negative. In either case, the potential of q would decrease.

CHAPTER 7 EVALUATION

MULTIPLE-CHOICE

1 D

2 B

3 C

SHORT ANSWER

4 $F = \frac{kQq}{r^2}$

$F = \frac{9 \times 10^9 \times 5 \times 10^{-6} \times 10 \times 10^{-6}}{0.05^2}$

$F = 180\,\text{N}$

5 $E = \frac{F}{q}$

$E = \frac{mg}{q}$

$E = \frac{3.6 \times 10^{-6}\,\text{kg} \times 9.81\,\text{ms}^{-2}}{2.0 \times 10^{-6}\,\text{C}}$

$E = 17.66\,\text{NC}^{-1}$

6

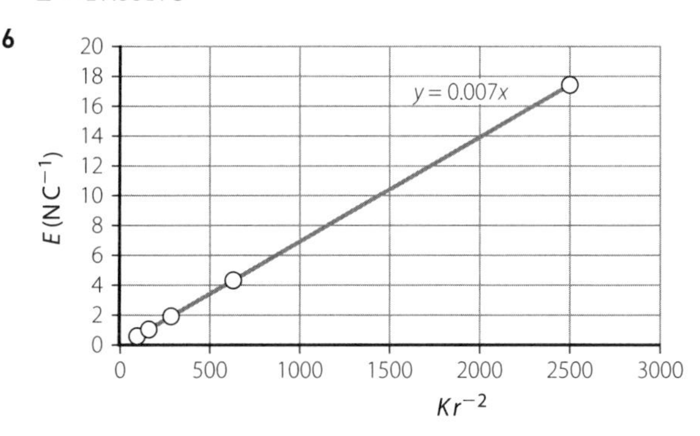

The point source emitting this field has a strength of 0.007 C

CHAPTER 8 REVISION

8.1 MAGNETIC PROPERTIES

1 All materials have some kind of reaction to magnets. Diamagnetic materials are weakly repelled by nearby magnets, paramagnetic materials are weakly attracted to nearby magnets and ferromagnetic materials are strongly attracted to nearby magnets.

2

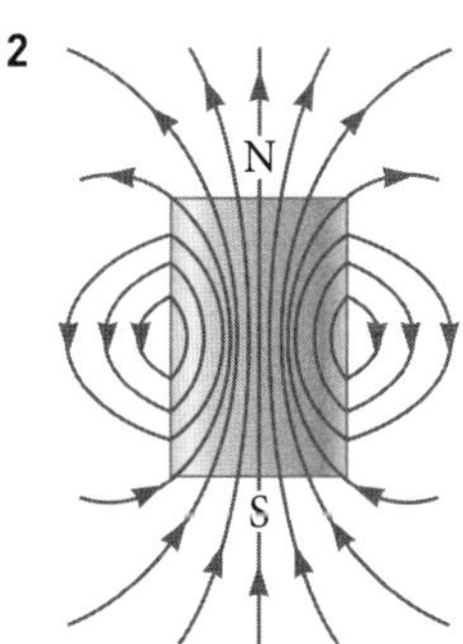

3 A stronger magnet can be represented by drawing more field lines around it.

4

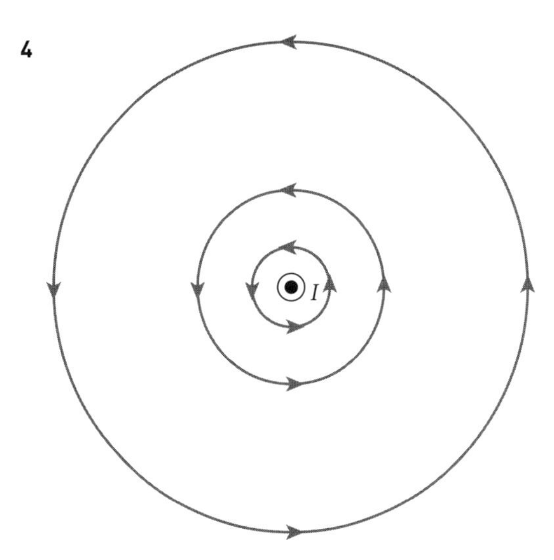

5 Iron is a ferromagnetic material and is strongly influenced by nearby magnets. When small, light iron filings are placed near a magnet, they will align with the magnetic field. This is because all the magnetic domains in iron will align with the field, causing the whole filing to do so as well.

8.2 MAGNETIC FIELDS DUE TO MOVING CHARGE

1 $B = \frac{\mu_0 I}{2\pi r}$

$7.0 \times 10^{-4}\,\text{T} = \frac{4\pi \times 10^{-7} \times 15\,\text{A}}{2\pi r}$

$r = \frac{4\pi \times 10^{-7} \times 15\,\text{A}}{2\pi \times 7.0 \times 10^{-4}}$

$r = 0.0021\,\text{m}$

2 $B = \frac{\mu_0 I}{2\pi r}$

$I = \frac{2\pi r B}{\mu_0}$

$I = \frac{2\pi \times 0.10 \times 8.9 \times 10^{-5}}{4\pi \times 10^{-7}}$

$I = 44.5\,\text{A}$

3 $B = \mu_0 n I$

$B = 4\pi \times 10^{-7} \times 200 \times 6$

$B = 1.5 \times 10^{-3}\,\text{T}$

4 $B = \mu_0 n I$

$n = \frac{B}{\mu_0 I}$

$n = \frac{6.6 \times 10^{-2}}{4\pi \times 10^{-7} \times 15}$

$n = 3501$ turns per metre

5 The centre of a solenoid is uniform. At the very edges, the field is not uniform as it is in concentric circles about each wire loop for every turn in the solenoid.

6 The bar magnet will either be repelled by the solenoid (two like ends of a magnet, such as N-N or S-S) or the bar magnet will be attracted into the hollow core of the solenoid.

7 900

8.3 FORCES ON CHARGES AND CURRENT CARRYING CONDUCTORS IN MAGNETIC FIELDS

1 A proton and electron differ in both the direction they move when entering a B field, and the radius of curvature of their path. If a proton has force acting on it to the left, the electron will have force acting on it to the right in the same B field. Since an electron is much less massive than the proton, its path will have a much larger radius of curvature.

2 $F = BIL\sin\theta$

$F = 5.0\times10^{-2}\,\text{T}\times3.0\,\text{A}\times0.5\,\text{m}\times\sin60°$

$F = 6.50\times10^{-2}\,\text{N}$

3 $F = BIL\sin\theta$

$B = \dfrac{F}{IL\sin\theta}$

$B = \dfrac{4\,\text{N}}{8.0\,\text{A}\times0.6\,\text{m}\times\sin0°}$

$B = 0.83\,\text{T}$

4 $F = qvB\sin\theta$

$B = \dfrac{F}{qv\sin\theta}$

$B = \dfrac{9.9\times10^{-11}\,\text{N}}{1.6\times10^{-19}\,\text{C}\times200\times10^{3}\times\sin0°}$

$B = 3094\,\text{T}$

5 In the second figure the electron would experience a force downwards initially, but then it would experience a force to the right once it is travelling downwards. After this, the electron would experience a force upwards and then finally a force to the left. This creates a circular path.

CHAPTER 8 EVALUATION

MULTIPLE-CHOICE

1 D

2 D

3 B

SHORT ANSWER

4 $F = qvB\sin\theta$

$F = 2\times1.6\times10^{-19}\,\text{C}\times4.8\times10^{4}\,\text{ms}^{-1}\times1.9\times10^{-2}\,\text{T}\times\sin0°$

$F = 2.9\times10^{-16}\,\text{N}$ upwards

5 $F = qvB\sin\theta$

$B = \dfrac{F}{qv\sin\theta}$

$B = \dfrac{2.5\times10^{-15}\,\text{N}}{1.6\times10^{-19}\,\text{C}\times v\times\sin45°}$

$B = \left(\dfrac{22097}{v}\right)T$ into the page

6 The B field will halve.

CHAPTER 9 REVISION

9.1 ELECTROMAGNETIC INDUCTION

1 a Electromagnetic induction – the production of an electromotive force (emf) or a voltage in an electrical conductor due to its dynamic interaction with a magnetic field

b Electromotive force – a difference in potential that tends to give rise to an electric current.

c Induced current – a current that is generated in a conductor due to an emf that has been induced through electromagnetic induction.

d Magnetic field – a magnetic effect of electric currents and magnetic materials.

e Magnetic flux – a measurement of the total magnetic field which passes through a given area.

f Magnetic flux density – the strength of a magnetic field per unit area.

2 $4.8\times10^{-3}\,\text{Wb}$

3 a 0

b $7.0\times10^{-3}\,\text{Wb}$

4 30°

9.2 FARADAY'S LAW OF INDUCTION

1

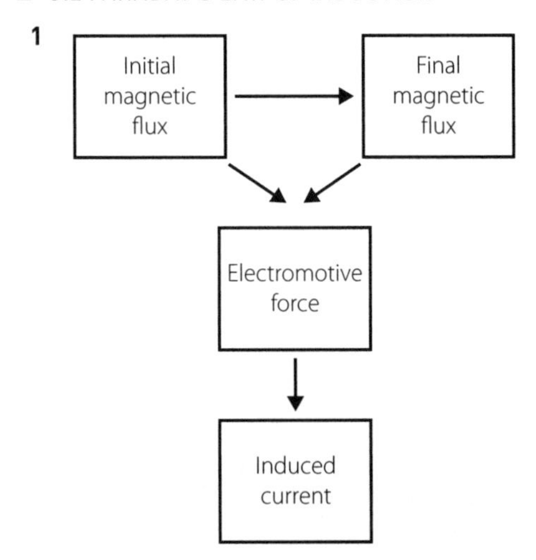

2 a Current is induced – rotating the wire will alter the portion of the magnetic field passing through the surface.

b Current is induced – the amount of magnetic field travelling through the surface is increased.

c No current is induced – the amount of magnetic field passing through the surface does not change.

d Current is induced – the amount of magnetic field passing through the surface will change.

9.3 SOLVING PROBLEMS USING FARADAY'S LAW

1 $1.1\times10^{-2}\,\text{V}$

2 27 A

3 25 A

9.4 LENZ'S LAW

1

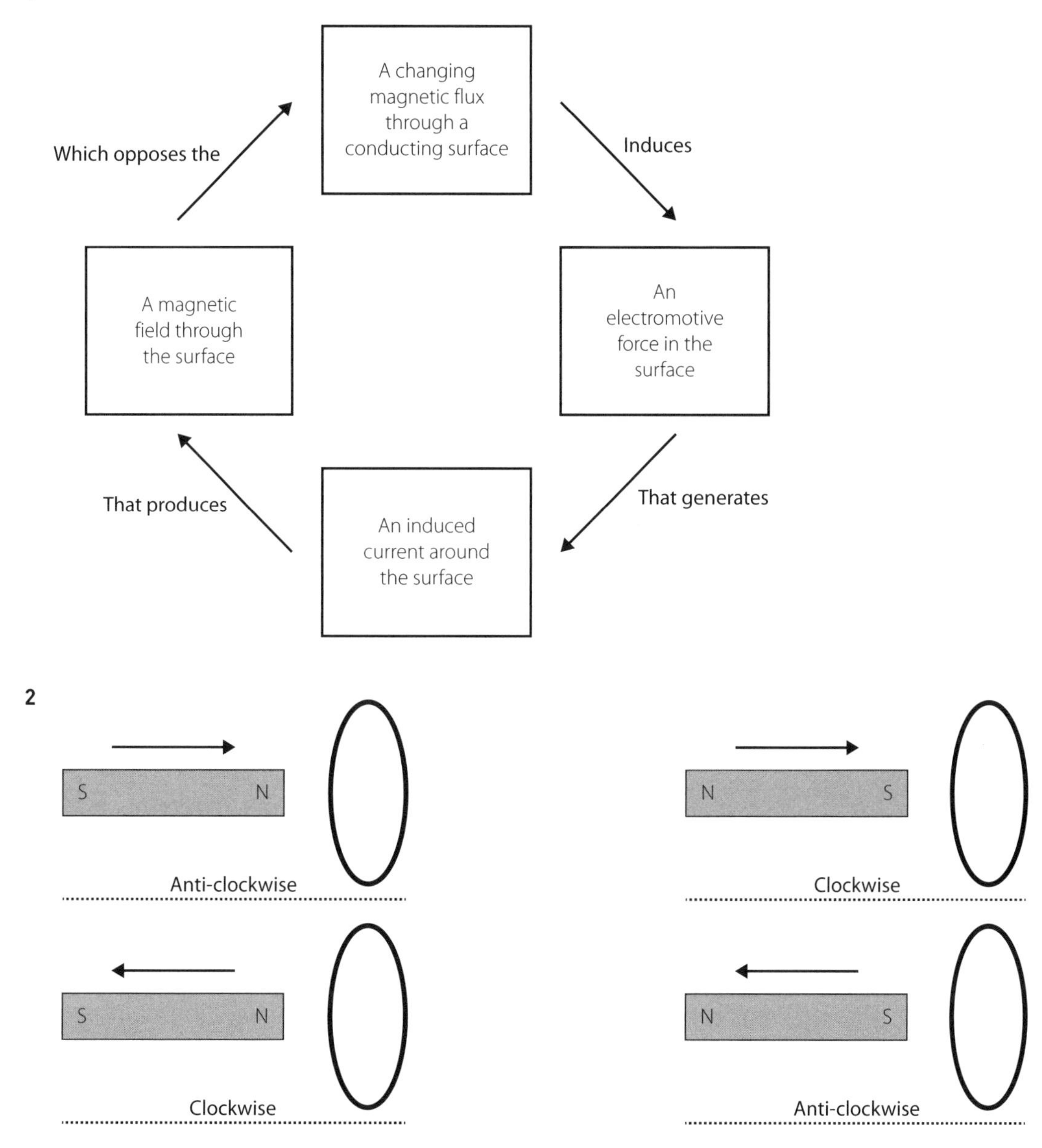

2

S N
Anti-clockwise

N S
Clockwise

S N
Clockwise

N S
Anti-clockwise

9.5 PRODUCTION AND TRANSMISSION OF ALTERNATING CURRENT

1

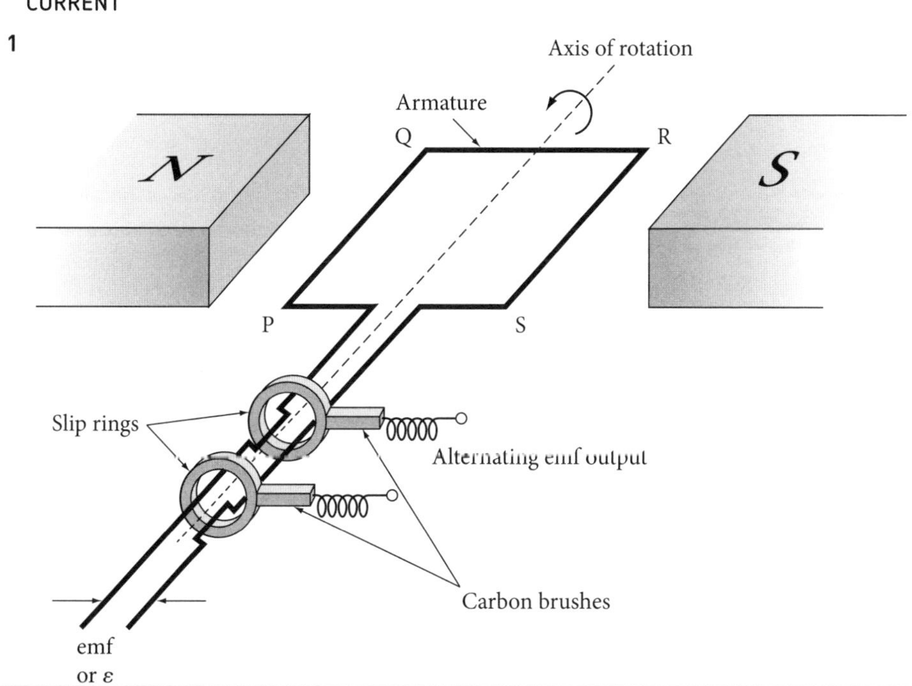

2 **a** $\Phi = 5\cos(100\pi t)$

b $\text{emf} = 500\pi\cos(100\pi t)$

c $V_{peak} = 500\pi V,\ V_{RMS} = \frac{500\pi}{\sqrt{2}} = 354\pi V$

d $I_{peak} = 200\pi A,\ I_{RMS} = \frac{200\pi}{\sqrt{2}} = 142\pi A$

e $P_{AV} = 5.0 \times 10^5\ W$

3 **a** $N_S = 7500$

b Step-up

c 2.0 A

CHAPTER 9 EVALUATION

MULTIPLE-CHOICE

1 A

2 C

3 D

4 C

5 C

SHORT ANSWER

6 Magnetic flux density

7 Opposite to the changing magnetic flux

8 Root mean square

9 Changing the magnetic field strength changes the magnetic flux which induces and emf in the conductor in accordance with Faraday's law. This emf will act upon any free charges in the conductor and cause a current to flow.

10 If the induced flux did not oppose the changing flux, then the current would produce a further increase in the magnetic flux through the loop. The flux would increase more, giving a bigger induced current, giving a bigger flux and so on.

11 84 mV

12 11 mV

13 **a** $\Phi = 9.375\cos(10\pi t)$

b $\text{emf} = 93.75\pi\cos(10\pi t)$

c $V_{peak} = 93.75\pi V$

d $V_{RMS} = \frac{93.75\pi}{\sqrt{2}} = 66.3\pi V$

e $P_{AV} = 8.67 \times 10^5\ W$

CHAPTER 10 REVISION

10.1 ELECTROMAGNETIC WAVES

1

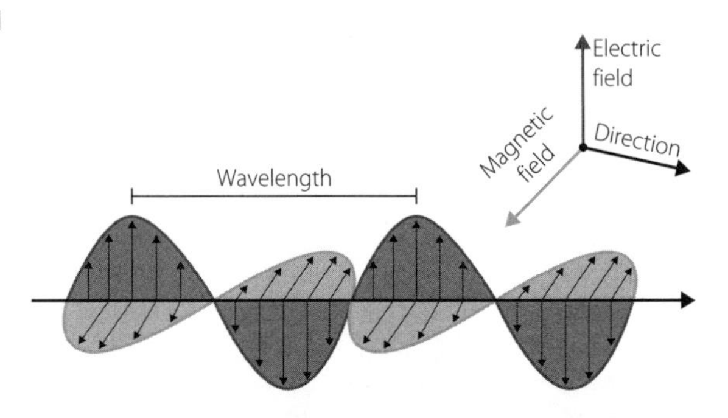

2 $1.26 \times 10^{-6}\ T\,m\,A^{-1}$

3 **a** X-rays

b $2.0\ 10^{-10}$ m

CHAPTER 10 EVALUATION

MULTIPLE-CHOICE

1 B

2 A

3 D

4 B

SHORT ANSWER

5 Transverse

6 Electromagnetic spectrum

7 Radio waves → microwaves → infrared radiation → visible light → UV rays → X-rays → gamma rays

8 An AC is used to produce an oscillating electric field, which by Maxwell's equations will create electromagnetic waves.

9 The oscillating electric field of an incoming electromagnetic wave causes the free charges in the receiving antenna to oscillate which is used to decode the information.

10 **a** Microwaves

b $3.7 \times 10^7\ m\,s^{-1}$

c 1.2×10^{-3} m

CHAPTER 11 REVISION

11.1 RELATIVE MOTION

1 Velocity of car with respect to person: $18\ m\,s^{-1}$, velocity of person with respect to car: $-18\ m\,s^{-1}$

2 $45\ m\,s^{-1}$ east

3 **a** Sunbather: $18\ m\,s^{-1}$ north

b passenger: $3\ m\,s^{-1}$ north

11.2 DEFINITIONS AND SIMULTANEITY

1 Author to supply answer

2 Author to supply answer

3 Paddy will see the fire-cracker at the rear of the cart flash first. As the car is moving away from Paddy at the moment the firecrackers go off, the light from the front sensor will need to travel further to reach him.

4 You will see the lightning strike tree B before A. As your friend is at the base of tree A, the light needs to travel 1 km from tree B to reach them. For you, the light only needs to travel 500 m, meaning you see the flash at B prior to seeing the flash at A sometime later. You also observe the flash at B before your friend sees the flash at B.

11.3 TIME DILATION, LENGTH CONTRACTION AND RELATIVISTIC MOMENTUM

1 On the satellite

2 The shorter distance will be measured by Amy who is travelling in the plane, assuming v is close to c.

3 $t = \frac{t_0}{\sqrt{1 - \frac{v^2}{c^2}}}$

$t_0 = t \times \sqrt{1 - \frac{v^2}{c^2}}$

$t_0 = t \times \sqrt{1 - \frac{v^2}{c^2}}$

$t_0 = 15\,\text{s} \times \sqrt{1 - \frac{(0.8c)^2}{c^2}}$

$t_0 = 15\,\text{s} \times 0.6$

$t_0 = 9\,\text{s}$

4 $m = \frac{m_0}{\sqrt{1 - \frac{v^2}{c^2}}}$

$m = \frac{1.67 \times 10^{-27}\,kg}{\sqrt{1 - \frac{(0.99999991c)^2}{c^2}}}$

$m = 3.93 \times 10^{-24}\,\text{kg}$

Now:

$p = mv$

$p = 3.93 \times 10^{-24}\,\text{kg} \times 0.99999991c$

$p = 1.18 \times 10^{-15}\,\text{Ns}$

5 $L = L_0\sqrt{1 - \frac{v^2}{c^2}}$

$L_0 = \frac{L}{\sqrt{1 - \frac{v^2}{c^2}}}$

$L_0 = \frac{45\,\text{m}}{\sqrt{1 - \frac{(0.6c)^2}{c^2}}}$

$L_0 = \frac{45\,\text{m}}{0.8}$

$L_0 = 56.25\,\text{m}$

6 $t = \frac{t_0}{\sqrt{1 - \frac{v^2}{c^2}}}$

$t^2 \times \left(1 - \frac{v^2}{c^2}\right) = t_0$

$1 - \frac{v^2}{c^2} = \frac{t_0}{t^2}$

$v^2 = c^2\left(1 - \frac{t_0}{t^2}\right)$

$v - \left(3 \times 10^8\,\text{ms}^{-1}\right)^2 \times \left(1 - \frac{3.0 \times 10^{-8}\,\text{s}}{\left(9.8 \times 10^{-8}\right)^2\,\text{s}}\right)$

$v = 2.91 \times 10^8\,\text{ms}^{-1}$

7 $p_v = \frac{p_0}{\sqrt{1 - \frac{v^2}{c^2}}}$

$p_v = \frac{1.67 \times 10^{-27}\,\text{kg} \times 0.95c}{\sqrt{1 - \frac{(0.95c)^2}{c^2}}}$

$p_v = \frac{4.76 \times 10^{-19}\,\text{Ns}}{0.3122}$

$p_v = 1.52 \times 10^{-18}\,\text{Ns}$

11.4 THE MASS–ENERGY EQUIVALENCY

1 $E = mc^2$

$E = 1\,\text{kg} \times (3 \times 10^8\,\text{ms}^{-1})^2$

$E = 9 \times 10^{16}\,\text{J}$

$1\,\text{kW hr} = 3.6 \times 10^6\,\text{J}$

So:

$\frac{9 \times 10^{16}\,\text{J}}{3.6 \times 10^6\,\text{J kW}^{-1}\,\text{hr}^{-1}} \times \$0.14 \text{ per kW hr}$

= \$3500000000

= \$3.5 billion

2 $E = mc^2$

$E = 9.11 \times 10^{-30}\,\text{J} \times \left(3 \times 10^8\right)^2$

$E = 8.2 \times 10^{-13}\,\text{J}$

3 $E = (\gamma m - m)c^2$

$E = \left(\sqrt{\frac{1}{1 - (0.95cc^{-1})^2}} \times 9.11 \times 10^{-31}\,\text{kg} - 9.11 \times 10^{-31}\,\text{kg}\right) \times \left(3.0 \times 10^8\,\text{ms}^{-1}\right)^2$

$E = \left(3.20 \times 9.11 \times 10^{-31}\,\text{kg} - 9.11 \times 10^{-31}\,\text{kg}\right) \times \left(3.0 \times 10^8\,\text{ms}^{-1}\right)^2$

$E = 1.81 \times 10^{-13}\,\text{J}$

4 The relativistic total energy of an object that is moving is its rest energy plus its relativistic kinetic energy. So, if the object is moving, then it can have more energy than just mc^2

CHAPTER 11 EVALUATION

MULTIPLE-CHOICE

1 C

2 A

3 C

SHORT ANSWER

4 $t = \frac{t_0}{\sqrt{1 - \frac{v^2}{c^2}}}$

$t^2 \times \left(1 - \frac{v^2}{c^2}\right) = t_0$

$$1-\frac{v^2}{c^2}=\frac{t_0}{t^2}$$

$$v^2=c^2\left(1-\frac{t_0}{t^2}\right)$$

$$v=\sqrt{\left(3\times10^8\right)^2\times\left(1-\frac{5}{10^2}\right)}$$

$$v=2.92\times10^8\ \mathrm{m\,s^{-1}}$$

5 Greater, because moving at relativistic speeds velocity is considered in the Lorentz transform of momentum.

CHAPTER 12 REVISION

12.1 QUANTUM THEORY KEY TERMS

1 Student's answers will vary

2 Student's answers will vary

12.2 THE NATURE OF LIGHT

1 The Photoelectric effect and black body radiation.

2 Visible light must be like an electromagnetic wave. The speed of light depends only on the medium.

3

SUPPORT FOR THE WAVE NATURE OF LIGHT	SUPPORT FOR THE PARTICLE NATURE OF LIGHT
Young's double slit experiment. Reflection, refraction, diffraction, wave interference.	Black body radiation. Photoelectric effect.

4

ELECTROMAGNETIC SPECTRUM REGION	APPROXIMATE WAVELENGTHS (m)
Gamma rays	$\times10^{-12}$
X-rays	$\times10^{-9}$ to $\times10^{-10}$
Ultraviolet light	$\times10^{-7}$
Visible light	$\times10^{-6}$ to $\times10^{-7}$
Infrared light	$\times10^{-5}$ to $\times10^{-6}$
Microwaves	$\times10^{-1}$ to $\times10^{-3}$
Radio waves	$\times10^{0}$ to $\times10^{2}$

12.3 YOUNG'S DOUBLE SLIT EXPERIMENT

1 Nodal points are points of destructive interference, where crests meet with troughs, creating a dark fringe. Antinodal points are points of constructive interference, where crests meet with crests, creating a bright fringe.

2 6.28×10^{-4} m

3 a Decreased

b Decreased

c Increased

4 6.71×10^{-4} m, 1.34×10^{-3} m, 2.01×10^{-3} m

5 a 2.08×10^{-6} m

b 4.99×10^{-2} m

12.4 WAVE PARTICLE DUALITY OF LIGHT

1 B

2 1.89×10^{-35} m

3 a 1.37×10^{-2} m

b 6.86×10^{-3} m

4 a $9.33\times10^{-24}\ \mathrm{kg\,m\,s^{-1}}$

b 4.14×10^{-11} m

c $\lambda_{Before}=7.10\times10^{-11}$ m

12.5 BLACK BODY RADIATION

1 A continuous spectrum represents radiation of all wavelengths over a given range, as for visible light.

An emission spectrum represents radiation emitted by an object such as a gas discharge tube.

An absorption spectrum represents the wavelengths of radiation absorbed by a material.

2 A black body is a perfect absorber. A black body also emits radiation with a spectrum that is characteristic of the temperature of the object.

3 $\lambda_{Max}=\frac{b}{T}$. T = 7155.6 K

4 Red, orange, yellow, white. $\lambda_{Max}=2.98\times10^{-6}$ m

12.6 PLANCK'S QUANTA AND PHOTON CHARACTERISTICS

1 $E=hf$; therefore, $h=\frac{E}{f}$ measured in Joules per hertz or Joules seconds (J s)

2 a 1.15×10^{-20} J

b 3.85×10^{-21} J

3 The emission spectrum is the spectrum of radiation emitted from a given object at a given temperature. Quanta describes a discrete unit of energy, charge, mass or other physical property, such as a photon. Energy levels are the allowed energies of a nucleus-electron system.

4 Electrons exist in quantised energy levels according to their orbital states. They can only take values in these discrete energy levels.

12.7 THE PHOTOELECTRIC EFFECT

1 A photoelectron is an electron ejected from the surface of a metal due to an incident photon.

2 a Silver

b Silver

3 a Gradient = 6.86×10^{-34} J s

b $W=hf_0=6.86\times10^{-34}\ \mathrm{J\,s}\times9.0\times10^{14}$ Hz

$W=6.17\times10^{-19}$ J

c Parallel to this line but, due to a higher work function, the line is slightly lower.

4 a Photoelectrons gain greater kinetic energy.

b Twice as many photoelectrons will be ejected.

5 a The number of photoelectrons represents the number of incident photons, hence the constant intensity.

b Varying frequency varies the kinetic energy available to photoelectrons as long as the threshold frequency is exceeded; therefore, D.

9780170412643

6

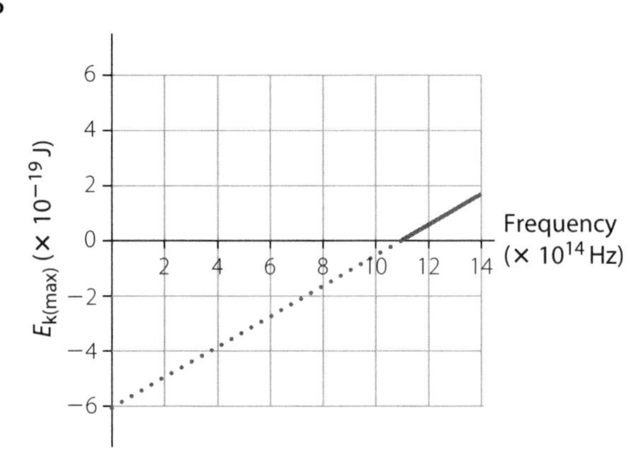

7 **a** 545 nm; green light

b $KE = 2.98 \times 10^{-19}$ J

8

WAVELENGTH (nm)	FREQUENCY ($\times 10^{14}$ Hz)	KE_{MAX} OF PHOTOELECTRONS (eV)	KE_{MAX} OF PHOTOELECTRONS (J)
647	4.64	0.74	1.18×10^{-19}
556	5.40	1.08	1.73×10^{-19}
490	6.12	1.49	2.38×10^{-19}
439	6.83	1.79	2.86×10^{-19}

a $h_{experimental} = 7.81 \times 10^{-34}$ J s

b $f_o = 3.14 \times 10^{14}$ Hz

c $W = hf_o = 2.45 \times 10^{-19}$ J

12.8 THE MODEL OF THE ATOM AND ATOMIC SPECTRA

1 Bohr's four postulates are as follows.

- An electron in an atom moves in a circular orbit about the nucleus under the influence of the electrostatic attraction of the nucleus.
- Only certain orbits are stable. Electrons in these orbits do not emit energy.
- The greater the radius of the orbit, the greater is its energy. Atoms emit radiation when an electron goes from one orbit to another orbit with lower energy. The energy released is: $E = E_f - E_i = hf$
- The orbits are characterised by quantised radii, given by: $r = \frac{nh}{2\pi m_e v}$. Where r is the radius in m, m_e is the mass of the electron in kg, v is its velocity, h is Planck's constant and n is an integer.

2 Ultraviolet light represents higher energy transitions than visible light. This atom may produce spectral lines in the ultra violet range from transitions to the n = 1 state.

3 $\lambda = 4.44 \times 10^{-11}$ m; $f = 6.76 \times 10^{18}$ Hz

4 $x = 10.86$ eV or 6.34 eV

5 **a** 10

b $n_5 \rightarrow n_1$

c $n_5 \rightarrow n_4$

CHAPTER 12 EVALUATION

MULTIPLE-CHOICE

1 D

2 C

3 C

SHORT ANSWER

4 7.63×10^{-7} m

5 9.35×10^{-7} m

6 **a** 4.71×10^{-2} m

b 2.36×10^{-2} m

7 A black body is an ideal surface that completely absorbs all wavelengths of electromagnetic radiation. A black body is also a perfect emitter of radiation.

8 0.168 m; 0.336 m

9 $E = hf = 2.82 \times 10^{-18}$ J; $W = 7.25 \times 10^{-19}$ J; $E > W$ hence photoelectrons will be ejected.

10 $\lambda = 4.04 \times 10^{-5}$ m. Increasing velocity decreases wavelength.

11 **a** 7.06×10^{12} Hz

b 4.68×10^{-21} J

c 0.029 eV

CHAPTER 13 REVISION

13.1 ELEMENTARY PARTICLES AND ANTIPARTICLES

1 **a** The ancient Greeks believed that the universe was made out of four fundamental elements: earth, air, fire and water.

b Democritus hypothesised that matter was composed of indivisible atoms, which were later shown to be composed of even more fundamental components including electrons, protons and neutrons.

c The standard model of particle physics uses experimental evidence from particle accelerators and cosmic rays to predict that these components were themselves made from smaller components termed elementary particles.

d An elementary particle is one which has no internal structure and cannot be divided into smaller components.

2 a Use the right-hand rule to determine how the force acts on the charged (positive) particle. The opposite occurs on the negative charged particle.

The force is directed at right angles to the motion, and produces circular motion. As the particle speed decreases the circular motion becomes a decreasing spiral.

b The positron decreased speed more quickly than the electron.

c The equation for centripetal force: $F = \frac{mv^2}{r}$

d The right hand rule relationship for current in a wire $F = BIL$ can be re-written. Where current times length can equate to charge times speed. For example, $IL = \frac{q}{t} \times L = q \times \frac{L}{t}$, where $\frac{L}{t} = v$ so that the force (F) on a positive charge (q) in a uniform magnetic field (B) becomes $F = qvB$.

e Combining Centripetal force relationship and force on a charge in a magnetic field relationship to find an expression for momentum ($p = mv$)

$$\frac{mv^2}{r} = qvB$$

$$mv = qBr$$

$$p = qBr$$

13.2 PARTICLE PHYSICS: THE CONTINUING SEARCH FOR ELEMENTARY PARTICLES

1 a Leptons

b Mesons

c Fermions, leptons and some baryons

d Bosons

e Leptons

f The delta particle

g Hadrons

2 a Electron

b Tau

c Kaon

d Eta or pion

e Omega

f Proton

13.3 GAUGE BOSONS AND THE FUNDAMENTAL FORCES OF NATURE

1 a A gauge boson is a force carrying particle that mediates particle interactions through the four fundamental forces.

b In the particle exchange model of interactions, one particle emits a gauge boson that is subsequently absorbed by another particle.

c The electromagnetic force is said to mediated by the photon.

d The strong nuclear force operates as a force of attraction between nucleons.

e The strong force operates as a force of attraction between quarks.

f The weak nuclear force is involved in nuclear decay and is mediated by W and Z bosons.

g The electroweak theory postulates that the electromagnetic and weak interactions are unified at high temperatures.

h The graviton is the gauge boson which has been hypothesised to mediate the gravitational force.

2

INTERACTION	RELATIVE STRENGTH	RANGE	GAUGE BOSON	MASS (GeV c^{-2})
Strong	1	1 fm	Gluon	0
Electromagnetic	10^{-2}	∞	Photon	0
Weak	10^{-5}	1 fm	W and Z bosons	80.4 – 91.2
Gravitational	10^{-39}	∞	Graviton	0 13.4 Leptons

9780170412643

13.4 LEPTONS

1

NAME	SYMBOL	ANTI-PARTICLE	MASS (MeV c^{-2})	B	L_0	L_U	L_T	LIFETIME (s)	SPIN
Electron	e^-	e^+	0.511	0	+1	0	0	Stable	½
Electron-neutrino	ν_e	$\bar{\nu}_e$	<7 eV c^{-2}	0	+1	0	0	Stable	½
Muon	μ^-	μ^+	105.7	0	0	+1	0	2.2×10^{-6}	½
Muon-neutrino	ν_μ	$\bar{\nu}_\mu$	<0.3	0	0	+1	0	Stable	½
Tau	τ	τ^+	1 784	0	0	0	+1	$<4 \times 10^{-13}$	½
Tau-neutrino	ν_τ	$\bar{\nu}_\tau$	<30	0	0	0	+1	Stable	½

13.5 HADRONS: MESONS, BARYONS AND THEIR QUARKS

1

NAME	SYMBOL	SPIN	CHARGE	BARYON NUMBER	STRANGENESS	CHARM	BOTTOMNESS	TOPNESS
Up	u	$\frac{1}{2}$	$+\frac{2}{3}$ e	$\frac{1}{3}$	0	0	0	0
Down	d	$\frac{1}{2}$	$-\frac{1}{3}$ e	$\frac{1}{3}$	0	0	0	0
Strange	s	$\frac{1}{2}$	$-\frac{1}{3}$ e	$\frac{1}{3}$	−1	0	0	0
Charmed	c	$\frac{1}{2}$	$+\frac{2}{3}$ e	$\frac{1}{3}$	0	+1	0	0
Bottom	b	$\frac{1}{2}$	$+\frac{1}{3}$ e	$\frac{1}{3}$	0	0	+1	0
Top	t	$\frac{1}{2}$	$+\frac{2}{3}$ e	$\frac{1}{3}$	0	0	0	+1

2

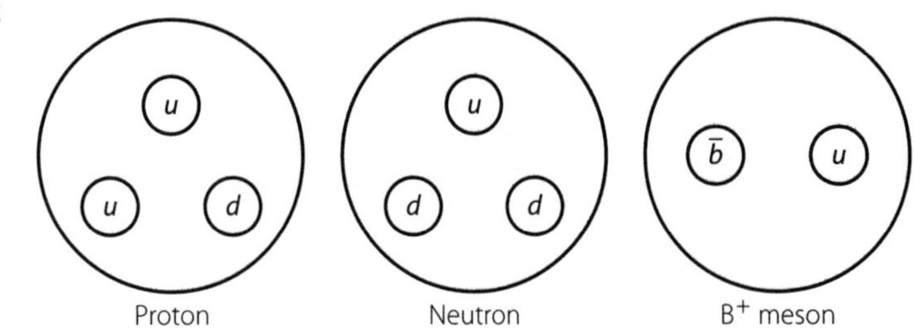

13.6 THE STANDARD MODEL TODAY

1

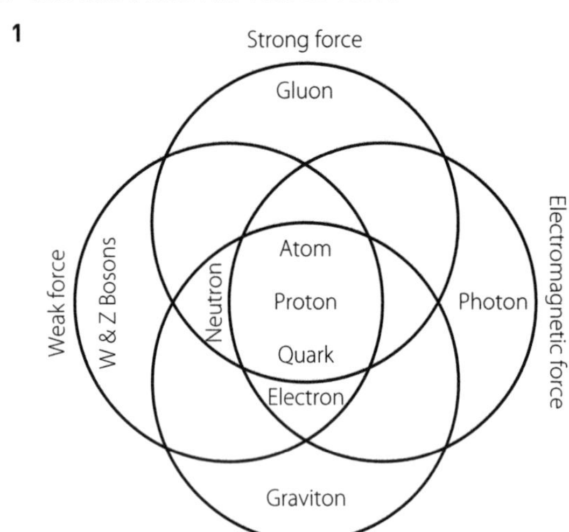

2

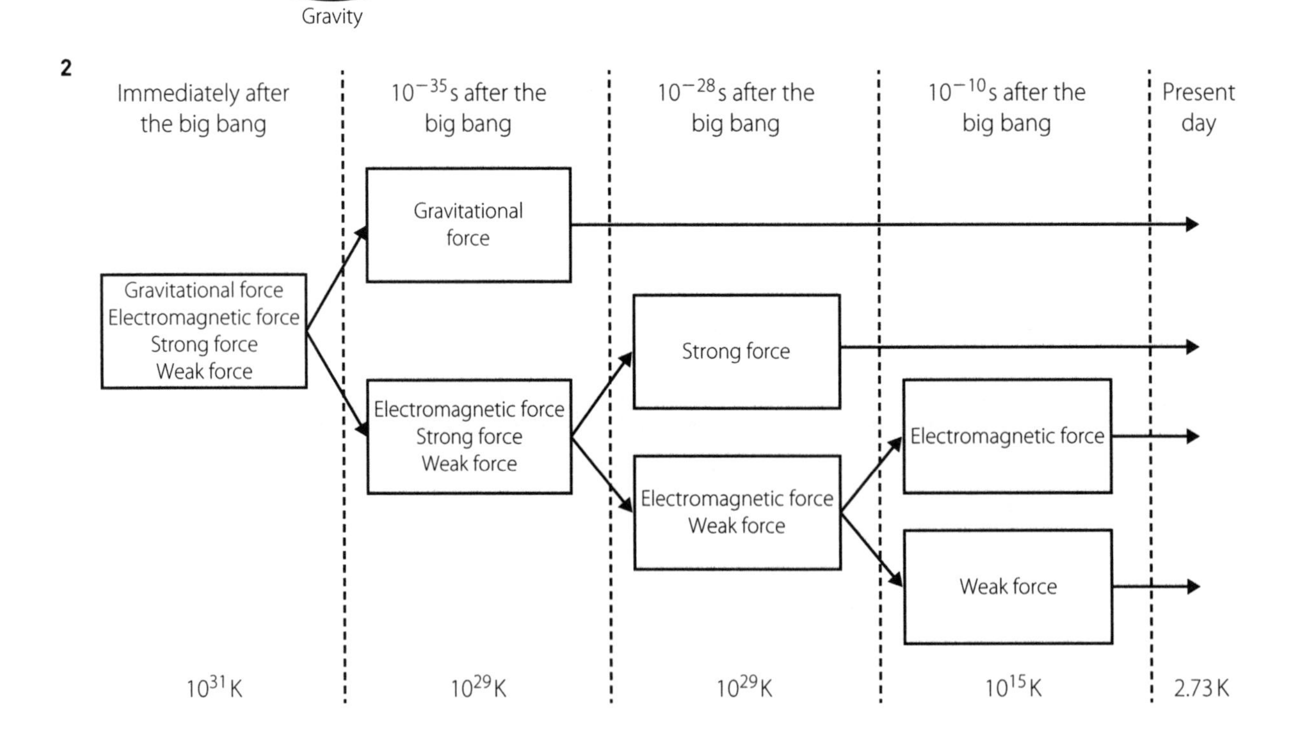

9780170412643

3

1 B	A	2 R	Y	O	N	N	3 U	M	B	E	4 R		5 P	
O		E					P				A		H	
S		D			6 S			7 N	E	U	T	R	O	N
O					P						E		T	
8 N	9 E	U	T	R	I	N	O						O	
	N				N			10 H	A	D	R	O	N	
	L							I						
11 D	A	R	K	E	N	12 E	R	G	Y		13 C			
	R					N		G			H			
	G					E		14 S	T	R	A	N	15 G	E
	I			16 Q		R		F			R		L	
	N			U		G		I			M		U	
	17 G	A	L	A	X	Y		E					O	
				R				18 L	E	P	19 T	O	N	S
				K				D			A			
20 I	O	N	I	S	E	D			21 E	Q	U	A	L	

CHAPTER 13 EVALUATION

MULTIPLE-CHOICE

1 B

2 D

3 C

4 B

5 C

6 D

SHORT ANSWER

7 Leptons, quarks and gauge bosons

8 Gravitational force and the graviton

9 Even though the electromagnetic and gravitational force have an infinite range, the gravitational force is approximately 10^{-39} times as strong as the electromagnetic force.

10 Inability to describe the gravitational force, prediction of magnetic monopoles and prediction that neutrinos should be massless.

11 Identity = proton, Spin = $+\frac{1}{2}$, Charge = +1 e, +1

12 a Electrostatic:

$$F = \frac{k_e q_e q_p}{r^2} = \frac{9 \times 10^{9}\,\text{N m}^2\,\text{C}^{-2} \times 1.062 \times 10^{-19}\,\text{C} \times 1.062 \times 10^{-19}\,\text{C}}{\left(5.3 \times 10^{-11}\,\text{m}\right)^2}$$

$F = 3.6 \times 10^{-8}$ N

Gravitational:

$$F = \frac{GMm}{r^2} = \frac{6.67 \times 10^{-11}\,\text{N m}^2\,\text{kg}^{-2} \times 1.67 \times 10^{-27}\,\text{kg} \times 9.11 \times 10^{-31}\,\text{kg}}{\left(5.3 \times 10^{-11}\,\text{m}\right)^2}$$

$F = 3.6 \times 10^{-47}$ N

b The electromagnetic force is 10^{39} times larger, very much stronger.

13 The strong nuclear force, holding the proton and neutron together.

CHAPTER 14 REVISION

14.1 CONSERVATION OF LEPTON AND BARYON NUMBER

1 Baryons are comprised of three quarks – leptons are not. Baryons interact with strong forces – leptons do not.

2 a Conservation of lepton number: 0 = 0, yes

Conservation of baryon number: 1 = 1, yes

b Conservation of lepton number: $1 \neq 3$, no

Conservation of baryon number: 0 = 0, yes

c Conservation of lepton number: $1 \neq 2$, no

Conservation of baryon number: 1 = 1, yes

d Conservation of lepton number: 0 = 0, yes

Conservation of baryon number: $1 \neq 2$, no

e Conservation of lepton number: 0 = 0, yes

Conservation of baryon number: 1 = 1, yes

14.2 REACTION DIAGRAMS AND FEYNMAN DIAGRAMS

1

n p e $\bar{\nu}_e$

2

Σ^- n π^-

3

e^- n p ν_e

4

Time e^- e^+ Space

5

Time n ν_e p e Space

6

Time ν_μ $\bar{\nu}_e$ e μ Space

14.3 SYMMETRY

1

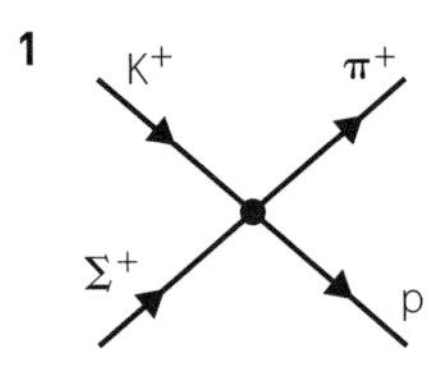

2 This is an example of anti-muon decay. There is not any reaction that is more likely to occur than another in this instance. Below if the following three possible crossing symmetries:

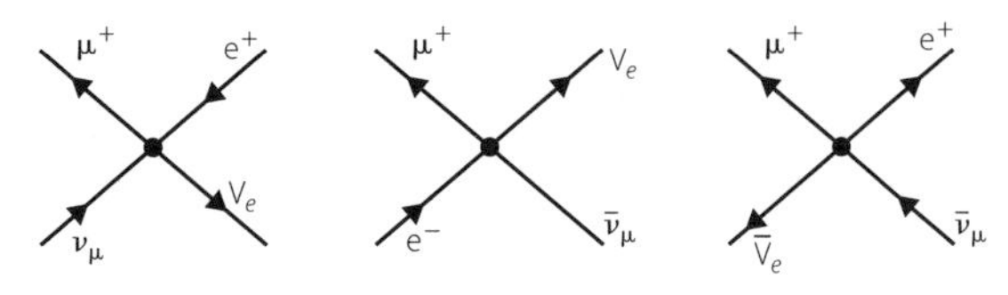

3

e^- $\overline{V}_e$ μ^- ν_μ

4 $n + \pi^- \rightarrow \Sigma^-$

5 $K^- \rightarrow \pi^+ + \pi^- + \pi^-$

6 $\tau^- + \overline{\nu}_\tau \rightarrow e^- + \overline{\nu}_e$

CHAPTER 14 EVALUATION

MULTIPLE-CHOICE

1 C

2 B

3 A

SHORT ANSWER

4 Yes, both lepton number and baryon number are conserved.

5

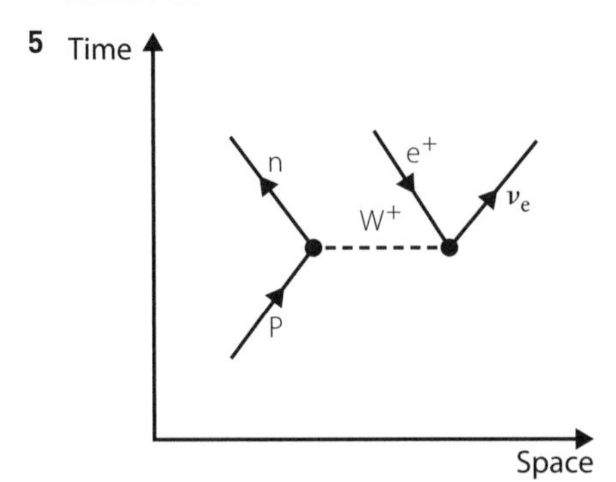

6 While the new reaction is always possible, sometimes the likelihood of having all the reactants in the same place is extremely unlikely, and hence not observable.

9780170412643

PHYSICS UNITS 3 & 4

Practice exam answers

MULTIPLE-CHOICE QUESTIONS

1 D
2 B
3 C
4 B
5 D
6 A
7 C
8 B
9 D
10 C
11 B
12 D
13 B
14 A
15 C
16 D
17 C
18 B
19 A
20 B

SHORT-RESPONSE QUESTIONS

1

LAUNCH VELOCITY ($m s^{-1}$)	LAUNCH ANGLE (°)	HORIZONTAL COMPONENT OF INITIAL VELOCITY ($m s^{-1}$)	VERTICAL COMPONENT OF INITIAL VELOCITY ($m s^{-1}$)	HORIZONTAL RANGE (m)
20	30	17.3	10	35.3
30	45	21.2	21.2	91.9
40	25	36.3	16.9	125.1
50	60	25	43.3	221.0

2 $8205\,m\,s^{-1}$

3 $8.60\,kg\,m\,s^{-1}$ west, $12.29\,kg\,m\,s^{-1}$ north

4 15.6 m

5 6.81 N

6 $9.16 \times 10^{6}\,s$

7 16.5 N, $2.1\,m\,s^{-2}$

8 2699 N

9 Force weight is 1176 N; the Fc required is 5269 N; the rider will not leave the ground.

10 0.87 s

11 $v = 0.952c$

12 **a** 0.047 m

b 0.024 m

13 Cosmic microwave background radiation

14 0.011 V

15 $6.09 \times 10^{-11}\,J$

16 $d = 6.28 \times 10^{-4}\,m$

17 $7.63 \times 10^{-7}\,m$ or 763 nm

18 B (+1) → (+1), L (0) → (+1) + (−1), yes

19 6.34 eV

20 Gluons; x-axis

COMBINATION-RESPONSE QUESTIONS

1

MOON	ORBITAL PERIOD, t (days)	ORBITAL RADIUS, r (m)
Io	1.78	4.21×10^{8}
Europa	3.56	6.69×10^{8}
Ganymede	7.10	1.06×10^{9}

2 **a** $5.85 \times 10^{9}\,J$

b $1.91 \times 10^{11}\,J$

c $1.97 \times 10^{11}\,J$

3 $T_{top} = 15.5\,N$, $T_{bottom} = 64.5\,N$

4 $5\,014\,m\,s^{-1}$

5 15 596 N

6 $N_S = 7500$ turns, step-up

7 n: neutron, p: proton, e: electron, $\bar{v}$: antineutrino. The conservation of baryon number (n, p) and conservation of lepton number (e, $\bar{v}$) is applied.

8 $W = 5.97 \times 10^{-19}\,J$

9

1. An electron in an atom moves in a circular orbit about the nucleus under the influence of the electrostatic attraction of the nucleus.
2. Only certain orbits are stable. Electrons in these orbits do not emit energy.
3. The greater the radius of the orbit, the greater is its energy. Atoms emit radiation when an electron goes from one orbit to another orbit with lower energy. The energy released is: $E = E_f - E_i = hf$
4. The orbits are characterised by quantised radii.

10

R (m)	R^2 (m^2)	$\frac{1}{R^2}$ (m^{-2})	E (NC^{-1})
0.02	0.0004	2500	19.25
0.04	0.0016	625	4.85
0.06	0.0036	278	2.17
0.08	0.0064	156	1.21
0.10	0.01	100	0.79

The gradient of an E vs $\frac{1}{R^2}$ graph is $7.69 \times 10^{-3}\,NC^{-1}\,m^2$.

$Q = 8.54 \times 10^{-13}\,C$